The American Poultry-book;

Being a practical Treatise on the Management of Domestic Poultry. By MICAJAH R. COCK. With illustrative Cuts. 18mo, Muslin, 35 cents.

American Husbandry.

Being a Series of Essays on Agriculture. With Additions, by WILLIS GAYLORD and LUTHER TUCKER. 2 vols. 18mo, half Sheep, $1 00.

French Domestic Cookery:

Combining Elegance with Economy; describing new Culinary Implements and Processes; the Management of the Table; Instructions for Carving; French, German, Polish, and Italian Cookery: in Twelve Hundred Receipts. Besides a variety of new Modes of keeping and storing Provisions, domestic Hints, &c. Management of Wines, &c. With Engravings. 12mo, Paper, 50 cents; Sheep, 62½ cents.

The Farmer's Dictionary.

A Vocabulary of the Technical Terms recently introduced into Agriculture and Horticulture from various Sciences, and also a Compendium of Practical Farming; the latter chiefly from the Works of the Rev. W. L. Rham, Loudon, Low, and Youatt, and the most eminent American Authors. Edited by D. P. GARDNER, M.D. With numerous Illustrations. 12mo, Muslin, $1 50; Sheep extra, $1 75.

The Trees of America,

Native and Foreign, Pictorially and Botanically Delineated, and Scientifically and Popularly Described; being considered principally with reference to their Geography and History; Soil and Situation; Propagation and Culture; Accidents and Diseases; Properties and Uses; Economy in the Arts; Introduction into Commerce; and their Application in Useful and Ornamental Plantations. By D. J. BROWNE. Illustrated by numerous Engravings. 8vo, Muslin, $5 00.

Chemistry applied to Agriculture.

By M. LE COMTE CHAPTAL. With a preliminary Chapter on the Organization, Structure, &c., of Plants, by Sir HUMPHREY DAVY. An Essay on the Use of Lime as a Manure, by M. PUVIS; with introductory Observations to the same, by Prof. RENWICK. Translated and edited by Rev. WILLIAM P. PAGE. 18mo, half Sheep, 50 cents.

Botany of the United States North of Virginia:

Comprising Descriptions of the Flowering and Fern-like Plants hitherto found in those States, arranged according to the Natural System. With a Synopsis of the Genera according to the Linnæan System, a Sketch of the Rudiments of Botany, &c. By LEWIS C. BECK, M.D. 12mo, Muslin, $1 25; Sheep extra, $1 50.

The Parlor Book of Flowers;

Or, Boudoir Botany. Comprising the History, Description, and Colored Engravings of 24 Exotic Flowers, 24 Wild Flowers of America, and 12 Trees with Fruits. With an Introduction to the Science of Botany. By J. B. Newman, M.D. Illustrated with 250 Engravings. 8vo, Morocco, gilt edges, $5 00.

The Cook's Oracle and Housekeeper's Manual:

Containing Receipts for Cookery and Directions for Carving. Also, the Art of composing the most simple and most highly finished Broths, Gravies, Soups, Sauces, store Sauces, and flavoring Essences; Pastry, Preserves, Pickles, &c. With a complete System of Cookery for Catholic Families. The Quantity of each Article is accurately stated by Weight and Measure; being the Result of actual Experiments instituted in the Kitchen of William Kitchiner, M.D. Adapted to the American Public by a Medical Gentleman. 12mo, Sheep extra, 87½ cents.

The American Shepherd.

Being a History of the Sheep, with their Breeds, Management, and Diseases. Illustrated with Portraits of different Breeds, Sheep Barns, Sheds, &c. By L. A. Morrell. With an Appendix, embracing upward of Twenty Letters from eminent Wool-growers and Sheep-fatteners of different States, detailing their respective Modes of Management. Illustrated with Engravings. 12mo, Paper, 75 cents; Muslin, 90 cents; Sheep extra, $1 00.

Domestic Duties;

Or, Instructions to Young Married Ladies on the Management of their Households, and the Regulation of their Conduct in the various Relations and Duties of Married Life. By Mrs. William Parkes. 12mo, half Sheep, 75 cents.

An Encyclopedia of Domestic Economy:

Comprising such Subjects as are most immediately connected with Housekeeping; as, the construction of Domestic Edifices, with the Modes of Warming, Ventilating, and Lighting them; a Description of the various Articles of Furniture; a general Account of the Animal and Vegetable Substances used as Food, and the Methods of Preserving and Preparing them by Cooking; making Bread; Materials employed in Dress and the Toilet; Business of the Laundry; Description of the various Wheel-Carriages; Preservation of Health, Domestic Medicines, &c. By Thomas Webster, F.G.S., assisted by the late Mrs. Parkes. Edited, with Notes and Improvements, by D. Meredith Reese, A.M., M.D. Illustrated with nearly 1000 Engravings. 8vo, Muslin, $3 50; Sheep extra, $3 75.

Dr. Oldcook's Receipt-book.

With Notes for the Preservation of Health. 12mo, Paper, 25 cents.

Queen Victoria's Poultry-House.—See Chapter V

THE

AMERICAN POULTERER'S COMPANION:

A Practical Treatise

ON THE

BREEDING, REARING, FATTENING, AND GENERAL MANAGEMENT OF THE VARIOUS SPECIES OF

DOMESTIC POULTRY,

WITH ILLUSTRATIONS, AND PORTRAITS OF FOWLS TAKEN FROM LIFE.

BY

C. N. BEMENT.

FIFTH EDITION.

NEW YORK:

HARPER & BROTHERS, PUBLISHERS,

329 & 331 PEARL STREET,

FRANKLIN SQUARE.

1852.

PREFACE

TO

THE FIFTH EDITION.

It is now about two years since the first edition of this work was published. Four editions have been taken off. The decided favor of the public, and an extensive patronage, seemed to indicate that a new edition was in request, and would, doubtless, be acceptable; it is now presented with some additions and improvements in its general appearance.

A few hints on the subject of rearing turkeys, furnished by a friend, on page 239, will, no doubt, be found very useful to the breeders of that valuable domestic fowl.

To the gentlemen of the press, who have so favorably noticed our humble endeavors, we tender our unfeigned thanks. Among the many testimonials of approbation, we take the following, being the first we can lay our hands on.

Mr. Breck, of the New England Farmer, thus notices the work:

"*Domestic Poultry.*—We have just received from the publishers a practical treatise on the breeding, rearing, fattening, and general management of the various species of domestic poultry, with illustrations and portraits of fowls taken from life. The author of the work is Mr. Bement, the popular landlord of the American Hotel, Albany. From long experience as a public caterer, Mr. Bement is fully qualified to judge of the qualities of poultry, and from his earliest youth, he says, he has taken great interest in rearing it as a matter of profitable investment. The subject is one which comes home to the bosoms and stomachs of every one, particularly about Thanksgiving time, and we would recommend this treatise to all practical farmers and agriculturists who may take an interest in the rearing of this kind of stock."

The editor of the Cultivator announces the work as follows: "*Amer-*

ican Poulterer's Companion.—This work is the most complete of the kind yet published in this country. It embraces 380 pages, 18mo. is beautifully printed, and contains many engravings, illustrating the characteristics of the different species and varieties, and exhibiting the most approved plans of buildings and other necessary appurtenances to the successful management of poultry. As a work of *practical* value, and containing much information on all branches of the subject to which it refers, we have no doubt it will be eagerly sought, and highly prized by the American public."

Mr. Allen, editor of the American Agriculturist, says, "We are glad at length to be able to announce this excellent work, which has been delayed some time passing through the press from the unusual care bestowed in engraving the illustrations. There are upward of seventy in number, and our readers can judge of their elegant finish, and general truthfulness and beauty, from the specimens given from it in our columns in the three preceding numbers of this journal. Mr. Bement has been a great poultry fancier for years, and has devoted much time to the breeding, rearing, and diseases of the different varieties of the domesticated feathered race. His book details his knowledge on this interesting subject, thus practically acquired; it may, therefore, be taken as a safe guide in all these matters, and more especially as it is adapted to our own country and its wants; thus making it emphatically *the* American Poulterer's Companion. Mr. Bement has treated the subject in a lively, agreeable style, and the publishers have issued it in the handsomest style of paper and typography. We are persuaded that the value of its contents, and general beauty of its appearance, will insure it a deserved popularity with all who take any interest in breeding and rearing poultry."

"*The American Poulterer's Companion.*—The value of this book is very decidedly indicated by the rapid sale of the first edition. We can not but add our testimony also to its value, after a pretty careful examination of its contents. It is the book which not only every farmer should possess, but also the mechanic, or every one who has a spot of land large enough to accommodate a dozen or two of fowls. In the publication of this work, Mr. Bement has certainly performed a very important service to the community, and the subject can not be considered a small one, when it is known that the value of poultry in New York alone amounted, according to the last census, to $2,373,029 and that in the states and territories it amounted to the sum of $12,176,170."—*Am. Qr. Jour. of Agriculture.*

"*The American Poulterer's Companion.*—A capital work this, and one which every person owning a single hen, and anxious to own more, should possess. It contains admirable lessons for the amateur and professional poulterer, and good will come from an extensive perusal of its practical pages."—*U. S. Gazette.*

The Philadelphia North American thus humorously notices the work. "*The American Poulterer's Companion.*—The author of this book is evidently no *chicken.* He takes up the subject *ab ovo*, and, from his study and experience, is *cock* sure of the correctness of all his facts and principles. The performance is a decided *feather* in his cap, and we hope that he may find the public ready to *shell* out in testimony of his success. It would be, to say the least, *foul* play if so much labor should not have its reward, and, so far as we are concerned, we feel it a duty to *spur* the attention of our country friends to the author's merits. Without meaning to tread on political ground, we are not sure but he has abundant reason to *crow.*"

CONTENTS.

CHAPTER IX. PAGE

CHAPTER X.

CHAPTER XI.

CHAPTER XII.

CHAPTER XIII.

CHAPTER XIV.

CHAPTER I.

INTRODUCTION.

Fowls reclaimed from their original state—breeding and rearing—some more prolific than others—considered a luxury—object of rearing for market—importance of rearing—mechanics may raise poultry—eggs and poultry prized by all nations—contribution of the poultry-yard profitable—utility of fowls—valuable as any stock—difference in richness of eggs—expensive establishments in England and Scotland—the way in which our farmers manage poultry not the best—every farmer may have an abundance of eggs—rearing poultry for market may be made profitable on small farms—importance—statistics.

THE feathered tenants of the farm-yard, reclaimed from their original state of dependence, and pensioners on our bounty, are peculiarly interesting. Though less decidedly important than the sheep, the pig, the cow or the horse, they still rank among the useful; their flesh and eggs are esteemed as wholesome and delicate food, and most are remarked for grace and beauty. In London, and other large towns in Europe, extensive markets for the sale of poultry are established, and the poulterer and the egg-salesman carry on a lucrative business. The general demand for the flesh and eggs of poultry acts beneficially upon the small farmer, and renders their rearing profitable.

"The breeding and rearing domestic poultry,"

1*

says Main, " as one of the branches of rural economy, includes two special though different objects. The first is that of rearing poultry for amusement and for the table of the owner; and the second is, doing the same thing with a view to profit."

Every poulterer and every farmer should be aware that some kinds of the domestic fowls are more prolific and hardy than others; that some are of much greater size, and that the flesh and eggs of some species or varieties are much superior in richness and flavor to others. The many suppose that a " pullet is a pullet, and an egg an egg, and that's an end of it;" not so, however, those gastronomes, the old Romans, according to Horace. The epicures were particular in the varieties of the fowls cooked, or that produced their eggs, and even went so far as to distinguish between eggs that were supposed to produce males and females, as the following couplet will show:

Long be your eggs far better than round,
Cock eggs they are, more nourishing and sound."

In this country, poultry has ever been considered a luxury, and consequently not reared in such considerable quantities as in France, Egypt, and some other countries, where it is used more as a necessary article of food, than as a delicacy for the sick, or a luxury for the table. " In France," says Mowbray, " poultry forms an important part of the live stock of the farmer, and it has been said of that country, the poultry-yards supply a much greater quantity of food to the gentleman, the wealthy tradesman and the substantial farmer, than the shambles do; and it is well known that in Egypt, it has been, from time immemorial, a considerable branch of rural economy, to raise domestic poultry for sale, hatched in ovens by artificial heat."

The object of rearing poultry and eggs for market, may appear to some but a small concern; but a glance at the late agricultural census, would surprise many who had paid little or no attention to the subject, or not been in the habit of reflecting on the various items that go to swell our agricultural prosperity.

The importance of rearing poultry, in a pecuniary point of view, has been little appreciated by the farmer, and on most farms very little attention is paid to the rearing and breeding a greater number than can subsist by picking up waste or refuse grain, or what might escape the pigs and be lost. They are considered as unprofitable, and a very insignificant part of live stock on the farm; still they should not be altogether neglected, for there are very few persons who do not like a *fresh-laid egg* or a *fine fat pullet;* and these are some of the fine things which happily can be had in good perfection by the farmer or mechanic, with very little trouble or expense.

In the advancing state of agriculture, a peculiar interest is, at the present moment, thrown around every means calculated to advance the interests of rural economy; domestic poultry, though last not least, now comes in for a share, and we are pleased to perceive that more attention has of late been directed to this subject. There is scarcely an agricultural paper which reaches us, that does not contain some inquiries, in regard to their management, varieties, properties, &c.

"He who adds," says Boswell, "to the productiveness of any object of nature, which can add a unit to the sum of human subsistence, and which can render that available for the purpose, which was wasted or useless before, must be deemed a benefactor to his species. In this light, even the rearing a few poultry may be viewed; for, by them, much of the refuse of the kitchen may again appear on the table in a new and better form; and if to them can be added the rabbit, the pig and the cow, there is no necessity that anything be lost or thrown away."

The industrious mechanic can easily associate the poultry-yard, to add to the comfort of his family, to render his leisure hours more profitable, and to convert his recreations into a reward. With proper arrangements and attention he may, either in a city or village, at a trifling expense keep at least twenty hens, that will furnish each year from ten to fifteen hundred eggs, and not far from one hundred chickens, plump and full grown, for the table.

Among all nations throughout the globe, eggs and poultry have long been used, and highly prized as articles of food. But, the lack of information or the bestowal of proper attention in the management of fowls, the small quantity and high price of eggs in our markets during the winter season, cause most persons, in moderate circumstances, to do without them, while those of larger means use them as expensive luxuries.

From our own experience we can safely say that there are few parts of the farmer's premises, that can be made to contribute, according to the amount of capital invested, more effectually to the comfort of the family, and if properly managed to the aggregate profit of the season, than the poultry-yard, and I am pleased to observe that more attention has of late been directed to the subject of domestic fowls. "Take care of the cents, and the dollars will take care of themselves," is an old maxim, and so far as the farmer's profits are concerned, I think a true one.

But few species of animals are of so much utility as the species of the fowl. Whether young, adult, old, male, or female, these birds afford light, wholesome and strengthening food, which is equally suited to those in good health, and to those in a sick or convalescent state; which the art of our modern epicures knows how to transform in a thousand different ways and always agreeable, but which is not less succulent when dressed with temperate plainness.

A writer in the Genesee Farmer says:—"Hens are useful, valuable, and as profitable as any stock on the farm; but, like other stock, they should have an enclosure by themselves at certain seasons of the year, especially in the spring, when the sowing and planting begins."

But, though most farmers keep fowls and raise their own eggs, there are many who have not learned the difference there is in the richness and flavor of eggs produced by fat and well-fed hens, and those from birds that have been half starved through our winters. There will be some difference in the size, but far more in the quality. The yolk of one would be large, fine colored and of good consistence, and the albumen or white clear and pure; while the contents of the other will be watery and meagre, as though there were not vitality or substance enough in the parent fowl to properly carry out and complete the work that nature had sketched. In order, therefore, to have good eggs, the fowls should be well-fed, and also provided, during the months they are unable to come to the ground, with a box containing an abundance of fine gravel, that they may be able to grind and prepare their food for digestion. Of eggs, those from the domestic hen are decidedly the best; but those of ducks and geese may be used for some of the purposes of domestic cookery.

At many of the country establishments in England, Scotland and Ireland, the buildings and yard for fowls are arranged on an extensive scale, comprising every necessary building, commodiously planned, and embracing every accessary required for the natural propensities, the comfort and protection of the various kinds. Apartments which can be occasionally heated for the tender birds; basins of water which can be frequently emptied and refilled, and several enclosures of grass or orchard grounds, as outlets for the poultry to range in alternately. The yards and outlets are also surrounded by high picket fences, to prevent the escape of the fowls or entrance of enemies. A keeper, male or female, is usually appointed to take care of the whole, and receive orders relative to the required supplies of the family.

In such establishments no expense is spared, either as to the quality or quantity of food necessary for their support; and, therefore, the various descriptions required by the cook are always of the best quality.

The way in which the farmers in general, in this country, manage their poultry is not the best for them or the fowls. They are allowed to run where they please, to lay and sit at any time they may deem expedient; when the hen comes off with her chickens, she is suffered to ramble about, exposing the young brood to cold and wet, which thins them off rapidly; no suitable accommodations are provided for their roosting-places, and they are allowed to find a place to roost where they can, probably in some exposed situation in a tree or out-house; no attention is given to feeding them; and, under such circumstances, it is not to be wondered at that few or no eggs are produced, that few or no chickens are raised, or that fowls are sickly or unprofitable.

When with so little expense to himself, a farmer may have an abundant supply of eggs, and raise one or two hundred chickens, it would seem strange that the poultry business should be so little attended to by the owners of the soil. Where crops are sown immediately around the barns, it may be inconve nient to have fowls run at large; but in many cases fifty or a hundred of these birds may be kept, not only without injury but with benefit. There are generally large quantities of grain scattered in the barn-yards, and lost unless eaten by fowls; there are myriads of insects, such as flies, bugs, worms, grass hoppers, &c., which require to have their numbers diminished by the cock and his followers; and, even if constantly kept up and fed, experience shows that, for the amount of capital invested, the poultry-yard contributes, in proportion, as great a return as any part of the farm.

Rearing poultry for the market near our cities is more or less carried on by those who have conveniences for so doing; and some keep a large number of hens for laying, but breed no chickens for sale; the eggs yielding much more certain profit.

On small farms, near cities and villages, the breeding and rearing of turkeys, ducks, and, in some instances, the keeping of geese, is found profitable.

To show the importance of paying more attention to this, though humble branch of the farmer's business, and that, however small it may appear to some, a knowledge of the amount consumed in some cities would astonish them.

STATISTICS.

The annual consumption of poultry and small game in the city of Paris usually amounts to 22,000,000 lbs.

"The quantity of eggs used annually in France exceeds," says one of the late journals," 7,250,000,000, of which enormous number Paris uses about 120,000,000."

"The importation of eggs from Ireland in 1837 to Liverpool and Bristol alone, amounted in value to £250,000. The importation from France the same year was probably greater."

"It appears, from the custom house returns of the year 1838, that eggs were imported into England (although loaded with heavy duties), from the continent to the value of more than a million of dollars."

"It appears," says M'Culloch, "from official statements that the eggs imported from France (into England) amount to about 60,000,000 a year; and supposing them to cost, on an average, 4*d.* per dozen, it follows that the people of the metropolis of Brighton (for it is to that place they are most all imported), pay £23,000 a year for eggs; and suppose the freight, importer's and retailer's profit, duty, &c., raise their price to the consumer to 10*d.* per dozen, their total cost would be £213,000."

The number of eggs imported into England from various parts of the continent, in 1839, was

83,745,723, and the gross amount of duty received for the same was £29,111.

It has been ascertained that half a million of eggs are consumed every month in the city of New York. One woman in Fulton market sold 175,000 eggs in ten weeks, supplying the Astor House each day with 1000 for five days of a week, and on Saturday, 2,500.

"When we look," says McQueen, "at the immense number of eggs brought from Ireland (50 tons of eggs, and 10 tons of live and dead poultry, having been shipped from Dublin alone in one day), and 66,000,000 of eggs imported from France to London alone; and this immense number, a trifle certainly to what are produced in this country (England), we shall cease to wonder at the large capital (£8,000,000) invested in poultry of all kinds. The quantity of eggs imported into Liverpool from Ireland, in 1832, was 4097 crates, value £81,940 sterling; which, at 6*d.* per dozen, gives 3,297,600 dozens of eggs, and the number 39,331,200. In 1833, the import had increased to 7,851 crates, or upwards of 70,000,000. The number imported into Glasgow from Ireland in 1835, by the custom house entries, was 19,321 crates, which, at nine eggs to the pound, gives the number 17,459,568."

It is stated in a Providence paper, that one sloop has regularly, for twenty-three years, made twenty-five trips a year from Westport, Mass., to that port, during which period she has carried to that market.

on an average, four hundred dozen of eggs each trip, making altogether a total of 3,450,000, averaging twelve and a half cents per dozen, amounting to $35,500. Large quantities of eggs have found their way from Ohio to our cities, by means of the canals and railroads. In May, 1842, seventy barrels, containing 70 dozen each, amounting in number to 58,800, were sent to Boston per railroad.

In December, 1793, the number of turkeys sent to London, by the stage coaches from Norwich alone, amounted to upwards of 2,500, weighing nearly 14 tons.

The week preceding Michaelmas day, 1830, forty tons of poultry were sent from Bury St. Edmund's, Suffolk, to London, 30 of which were geese; and 16 tons of the latter were the property of Messrs. Flatt & Walton, poulterers of Yostock and Repworth. Mr. Clarke, of Boston, transmitted to London in December, 1833, the following quantity of poultry:—2400 geese and 800 turkeys; Mr. Harris, poultry-man of Spalding, also killed and forwarded to Leadenhall market, 1150 geese, 500 turkeys, 200 ducks, and 30 dozen of fowls.—*Mowbray.*

The production and consumption of poultry and game in Europe, may be judged of by the consumption of Paris, in 1832, which comprised the following articles and animals, according to Count Chalsol:—931,000 pigeons, 1,289,000 chickens, 549,000 turkeys, 238,000 geese, 131,000 partridges, 177,000 rabbits, and 174,000 ducks.

The editor of the Farmer's Cabinet, published at Philadelphia, says, " A farmer, who regularly attends market, sold during one year poultry and eggs to the amount of about $150, and the expense incurred in their production was so small as scarcely to be appreciable."

By referring to the agricultural statistics of the United States, as furnished by the last census, taken in 1839, and published in 1840, it would appear the value of poultry in the State of New York amounted to $2,373,029, and that of the various States and Territories of the Union amounted to the sum of $12,176,170.

More attention has been directed to poultry in the vicinity of Philadelphia, than any other part of our country. The Bucks county poultry, like the Dorking of Surrey, in England, have acquired a greater degree of celebrity even in the New York market, where they are sometimes found in abundance.

The foregoing is proof of the magnitude of an interest, which is deemed by most farmers of too trifling consideration to be worth making any calculation about. It is, without doubt, a proportionately great interest in this country; yet, who in this respect deems it worth attending to?

In England there are exhibitions and prizes awarded for the best poultry. At their exhibitions noblemen of the highest rank become competitors. Earl Spencer, in 1837, carried the highest prize for a turkey, which weighed 20¼ lbs. One capon was exhibited which weighed 7 lbs. 14 oz one pullet 6 lbs. 3 oz.; one goose 18 lbs. 2 oz.; one pair of ducks, 10 lbs. 10 oz.

Here in America, without the aid and stimulus of exhibitions and prizes, turkeys have been known to weigh over 30 lbs. In a Philadelphia paper it is stated that, "On Thursday, December 29th, 1842, a farmer from New Jersey obtained $\$10\frac{1}{2}$ for a turkey curiously. The farmer boasted that his turkey weighed 30 lbs., and asked a price for it proportioned to its dimensions. A customer, doubting this, said he would give him five dollars for the turkey if it weighed 25 lbs., and one dollar for every pound over that weight. The defunct turkey was put in the scales and weighed $30\frac{1}{2}$ lbs. The gentleman kept his word, paid $\$10\frac{1}{2}$, and took his fowl home for his New Year's dinner."

A pair of chickens were exhibited at one of the hotels in Philadelphia, in April, that weighed $19\frac{1}{4}$ lbs. after being dressed. These fowls were bred and fattened by the Messrs. Woods of Haddonfield, N. J.

In December, 1822, two turkeys were bred and fed, and sent to Cork, one weighing 33, the other 34 lbs.

In the 7th edition of Mowbray on Poultry, it is asserted that "three turkeys were sold at Leadenhall market, December 25th, 1833, which weighed together 91 lbs., and brought three guineas each. One, eighteen months old, weighing 34 pounds, was sold at the same price."

CHAPTER II.

GENERAL VIEWS: DOMESTIC POULTRY.

Propagation and crossing—acquaintance of nature and habits—the cock a living clock—life and attitudes of the cock—mythology, &c.—begins to pay his addresses—peace does not long exist—disposition for fighting—horrible story of Ardesoif—choice of the cock—cleanliness—plumage—poultry-yard a place for the study of natural history—change of colors—best soils—should be kept confined—care entrusted to competent persons—will become too fat—difference in well-fed fowls—easy to feed.

UNDER the term Domestic Poultry, are understood the cock and hen, turkey, duck, goose, pea and guinea fowl, to which perhaps may be added, the swan. Although fowls used for the table are, by nature, granivorous, yet all the various species, the goose perhaps excepted, are carnivorous likewise, and great devourers of fish and flesh.

By propagation and crossing, gallinaceous fowls have been distributed into endless variety; but without including the more marked breeds, Dr. Bechstein distinguishes eight varieties of the common barndoor fowl; viz. the fowl with a small comb; the crowned fowl; the silver-colored fowl; the slate-blue fowl; the chamois-colored fowl; the ermine-like fowl; the widow, which has white tear-like spots on a dark ground; and the fire and stone-colored fowls.

It is difficult, however, in many cases to identify the distinctions mentioned by foreign writers with the fowls bred in this country.

If one wishes to be acquainted with the nature and the inclinations of fowls, one is obliged to have recourse to the poultry-yard; for we know nothing of the habits of wild fowls; but a long bondage has operated such great alterations in the nature of our fowls, that it is not easy to come at their original character. For instance, the tame fowl makes no nest; the wild one surely does. The fecundity of the former is in a measure unbounded; except in the moulting season, it lays almost incessantly; analogy will not allow us to doubt but that, in the wild tribe, the laying must be considerably confined, and that it takes place only at regular times.

The cock is to the farmer a living clock, where exactness, to be sure, is not quite as correct as some of our Connecticut made wooden clocks; but is sufficient, nevertheless, to point out the divisions of the day and night, of labor and rest.

The attitudes of the cock are those of haughtiness; he carries his head high; his look is bold and quick; his gait is grave; all his motions bespeak a noble assurance; he seems to reign over the other inhabitants of the poultry yard. His activity is indefatigable, and he is never deficient in vigilance. Incessantly taken up with his mates, he warns them out of danger, gets before them, and if obliged to yield to force, which robs him of one, he for a long time expresses by loud outcries, his anger and his regrets; feeling for their suffering, he again utters long and sonorous exclamations, when by their cries they announce the pains or fatigues of laying. A softer clucking is the signal by which he calls them; his usual shrill crow is, at the same time, the expression of his continual vigilance; the cry of victory after an engagement, and the accent of satisfied love. It was formerly thought that the cock and the nightingale were the only day birds that sung and crowed at night; other species also warble after sunset; but all, as well as the nightingale, are quiet when the season of love is over; whereas the tame cock crows every day and every night throughout its whole existence. However, there is some ground to presume, that it is otherwise in a state of nature, and that the crowing of the wild cock is no more, as with other birds, than the momentary accent of his loves.

If the life of the domestic cock be an uninterrupted series of enjoyments, it is also commonly a continual scene of war. As soon as a rival comes forward, the fight begins, and only ends by the retreat of one of the champions. Sometimes both rivals die in the battle. If one of them be conqueror, he immediately celebrates his triumph by repeated crowings and by flapping his wings. The other disappears, abashed at being defeated.—*Buffon.*

"Less spirited than the males, hens are also milder and more timid; though they fight with each other, and for a moment, with ten times more fury than the cocks. Their voice is less sonorous; but its different modulations show that they, as well as cocks, have a varied language; after having laid, they utter loud cries; if they call their chickens together, it is by a short grave clucking; they warn them out of danger by a monotonous and lengthened cry, which they repeat till the bird of prey is out of sight; in fine, they keep up, between themselves, a continual cackling, which seems to be a coherent conversation between these very chattering females. There are some hens which faintly imitate the crowing of the cock; they are usually the young ones of the year, and they do not always keep on this mimic fancy, as I have ascertained by following several of those crowing hens, which happened to be at different times in my poultry-yard. As to the rest, they had none of those exterior characters which could bring them near the cock; they lay like the rest, and it is wrong that they should be generally proscribed, as either barren or as ill omened. The housewives of Lorraine, and several other parts of France, are forward in putting to death every hen that imitates the crowing of the cock, which in their eyes is the effect of a charm; hence a very jocular saying, in which there is some meaning, '*a hen that crows, a parson that dances, a woman that talks Latin; never come to any good.*'

"In the mythology of the ancients," says Main, "the cock was the symbol of vigilance. Polytheism consecrated it to Minerva and Mercury: it was offered to Æsculapius, the God of medicine, on recovering from illness. The Romans used to keep sacred pullets, and they undertook nothing of consequence before they had consulted the auspices of this prophetic fowl. Its meals were solemn omens, which regulated the conduct of the senate and the armies."

The cock is remarkable for his haughty, grave, stately gait, for his courage and vigilance, for his attachment to his hens, for his amorous disposition, and his means of satisfying it.

The cock begins to pay his addresses to the hens from the time he is four months old; his full vigor only lasts three years, though he may live till ten. It is remarked that in cocks of the large species, the procreative qualities are later in coming forward; they probably enjoy it longer. As soon as the cock gets less nimble he is no more worthy to figure in the seraglio; his successor must be the finest, the most brave of all the supernumerary young cocks in the poultry-yard.

Peace does not last long between cocks, among which the empire of the poultry-yard has been divided; as they are all actuated by a restless, jealous, hasty, fiery, ardent disposition, their quarrels are frequent, and are generally bloody. A fight soon follows the provocation. The two adversaries face each other; their feathers are bristled up, the neck stretches out, the head low, the bill ready; they observe each other in silence, with fixed and sparkling eyes. On the least motion of either they set off together, they stand stiff, rush forward, and dash against each other, and repeat the same manœuvre, till the one that is most adroit, and is strongest, has torn the comb of his enemy, has thrown him down, by flapping him with his wings, or has stabbed him with his spurs.

This disposition of cocks for fighting so desperately, especially when they are not used to live together, and meet for the first time, the courage and obstinacy which they evince in this often dreadful contest, have given Englishmen the idea of exhibiting these cock fights in public. It is that sort of tragedy they seem to like in preference. The annals of these sights mention a very singular sympathy between two cocks. They had successfully beaten all the others; they could never be made to fight together, notwithstanding the stimulus of the most hateful passions.

Mowbray relates the following: " Every one has heard the horrible story of Ardesoif of Tottenham, who, in April, being disappointed by a famous game-cock refusing to fight, was incited by his savage passion to roast the bird alive, whilst entertaining his friends. The company, alarmed by the dreadful shrieks of the victim, interfered, but were resisted by Ardesoif, who threatened death to any who should oppose him; and in a storm of raging and vindictive delirium, and uttering the most horrid imprecations, he dropped down dead. I had hoped to find this one among the thousand fanatical lies which have been coined on the insane expectation that truth can be advanced by the propagation of falsehood; but to my sorrowful disappointment, on a late inquiry among the friends of the deceased miscreant, I found the truth of the horrible story but too probable."

The choice of a cock is a very important thing. It is accounted that he has every requisite quality, when he is of a good size, but middling, when he carries his head high, has a quick and animated look, a strong and shrill voice, the bill thick and short, the comb of a fine red, and in a manner varnished; a membraneous wattle of a large size, and colored the same as the comb, the breast broad, the wings strong, the plumage black, or of an obscure red, the thighs very muscular, the legs thick, and supplied with long spurs, the claws supplied with nails rather bent, and with a very keen point; when he is free in his motions, crows often, and scratches the earth with constancy, in search of worms, not so much for himself as his mates; when he is brisk, spirited, ardent, and clever in caressing them, quick in defending them attentive in soliciting them to eat, in keeping them together in the day, and assembling them at night.

There are some cocks, which, by being too high mettled, are snappish and quarrelsome. The way to quiet these turbulent ones is plain; their foot must be put through the middle of a bit of leather in a round shape; they become as quiet as men who are fettered at their hands, feet, and neck.

The cock loves cleanliness; he is careful of his

coat; you often see him busy in combing, polishing, and stroking his feathers with his bill. If, like the robin and thrush, he has not the ambition of excelling in his note, one may at least think that he is particularly jealous in showing that he has a very loud, shrill, and powerful voice. In fact, when he has crowed, he listens to know whether he is answered; or, should he hear another, he begins again directly, and he seems to defy him to raise his voice above his own. Often of a dark night, this crowing, repeated by every cock in the village, has fortunately reached the ear of the benighted traveller, and has enabled him the better to direct his steps.

The plumage of birds has always formed an object of pleasing contemplation. The God of nature has shown by it his love of spreading beauty over all his works, and opening up every source for the pure enjoyment of man. The splendid coloring of many of our domestic fowls is not necessary in itself, and must have been bestowed as a means of pleasure to the beholder.

" If people," says M. Reaumur, " are affected with the kind of pleasure so transitory to the most enthusiastic florists, who procure it but for a few days, by a world of care and toil, continued through a whole year; if they are affected by the variety and fine combinations of colors in their favorite flowers, the poultry-yard, when well managed, may be made to offer them endless pleasures of the same description."

The greater number of cocks, even of the most

common kind, are beautifully pencilled, and when exposed to the play of the sun's rays exhibit the brightest hues, almost rivalling the gorgeous coloring of the rainbow. The hens are sometimes spotted with great beauty and regularity; some white and silvery, others by their bright orange tints appearing golden, while of the most common kinds there is an almost endless variety. In their colors they embrace the opposite extremes of light and shade, and all the tints that lie between them.

These colors are sometimes submitted to very remarkable changes in the same individual, at different stages of their existence. When newly hatched, the acutest poulterer could not predict of what precise color they would become, for it is not found invariably to run in the blood. After moulting, some fowls have been known to turn out a different color from what they were before. Even without moulting the feathers of the white have been tipped with black as suddenly as the hair of some men has in the course of a night been turned into grey.

The poultry-yard is a place where the student of Natural History will see many things to amuse and instruct. The changes of color which some of the domestic fowls undergo in the process of moulting, are most singular and inexplicable. M. Reaumur gives the following instance of change of color, among many others. "One of his hens, readily distinguished by a crooked claw, had feathers of the ruddy color mixed with brown, so common among dung-hill fowls. A year after, she was observed to become almost black, with here and there a large white spot. At the second moulting, black was the predominant color, and only a few white patches of the size of a half-crown could be perceived. At the succeeding moult, all the black disappeared, and the hen became pure white." In another case of a cock presented to M. Reaumur as a curiosity, the following changes occurred. In the first year he was of the common ruddy brown mixed with white; in the second, he was all over ruddy brown, or rather red without white; in the third, uniformly black; in the fourth, uniformly white; and in the fifth, white feathers mixed with chestnut and ruddy brown; while at the next moulting he again became a pure white.

A similar case lately occurred within the knowledge of the author. Passing a neighbor's yard in the month of July, I observed a beautiful young cock of the Poland variety. His color was red and black, beautifully combined, with a splendid white top-knot of feathers. Wishing to obtain him, I called there in January following, and on inquiry, he was produced to me perfectly white; I objected to him, observing to the owner, that it was a speckled fowl that I wished—one which I saw there in the summer. I was then informed that he was the identical fowl, and that he was the only cock which had been on the premises, and that when he moulted in the fall his color changed by degrees until every dark feather disappeared.

The following curious circumstance, which happened within the memory of many of the inhabitants of, and near Bath, is well worth stating, respecting poultry changing their plumage. Major Brereton, of the above place, had a noted game-cock, entirely of a dark red; and, after his great match, on which depended the sum of £36,000, and winning the odd battle, he turned him to a *walk*, at a place near Bath; the bird had not been long there, when the owner of the farm came to the Major, and informed him, he was all spangled with white; in a few days after, when the Major went to see him, he found him all over white; or, as it is termed by cockers, *a complete smock*—not a red feather was to be seen. In the course of some time after, he resumed his former plumage.

Dickson, in his work on poultry, with regard to color, relates the following: "I have, at present, a hen of the Spanish breed, which has been of a uniform black for two successive moults, but has now her neck, wings, and tail feathers, tipped with pure white. I have another which was all over a silver grey, but has now her head and neck coal black, with a ring of fine white at the base of the neck, while the rest of the body is finely speckled with black and snow white. It is remarkable also, that this change took place in a few weeks, without any obvious moult, so as to cause her to appear anywhere bare of feathers."

Mowbray, in his work on poultry, relates the following instance of the change in color of a turkey. " A turkey cock, the property of J. Lee, Esq., which was black in 1821, became afterwards *perfectly* white. This extraordinary change took place so gradually, that in the middle of the moulting, the bird was beautifully mottled, the feathers being black and white alternately."

The warmest and driest soils are best adapted for the purpose of breeding and rearing poultry. The greatest success may be expected, attended with the least trouble ; however, those who choose to keep them, must use the best place they can command. If possible, it must be a gentle slope, that the water and damp may run off.

In both large and small establishments, it will be required to separate some of the fowls from the rest, particularly when particular breeds, such as Dorking, Poland, Malay, or any fancy fowls, are to be reared. Separate pens or apartments must be provided, either at some distance from each other, which is preferred, or with divisions to prevent any intrusion, by which improper crossing might be produced.

It is well said by Mr. Beatson, that poultry, when rightly managed, might be a source of great profit to the farmer; but where many are kept, they ought not to be allowed to go at large, in which case little or no profit can be expected; for not only many of their eggs will be lost, and many of themselves perhaps destroyed by vermin, but, at certain seasons, they do much mischief both in the barn-yard and in the field. Poultry, it is thought, ought always to be confined; but if so, instead of a close, dark, diminutive hovel, as is often the case, they should have a spacious, airy place, properly constructed for them. And, from my own experience, each sort of poultry should be kept by itself. Though I have found turkeys and barn-yard fowls very quiet through the day, they do not like to roost together under the same roof; the turkeys are quarrelsome, and will not suffer the hens to come near them.

It is best to entrust the management of fowls to some trusty person, who can be depended on; and no other person, except the keeper, whom the fowls know, and the voice and sight of whom rejoice them, must go into the hen-house, for fear of scaring or disturbing the hens busied in laying.

The proper persons, or those who generally understand the art of rearing poultry, are females, who, accustomed from their infancy to look after the poultry, are acquainted with every particular of rearing, the different processes it requires, and the alterations which circumstances compel to bring forward.

Fowls will become fat on the common run of the barn-yard, where they thrive upon the offals of the stable and other refuse, with perhaps some small regular daily feeds; but, at threshing time, they become particularly fat, and are therefore styled barn-door fowls, probably the most delicate and high-flavored of all others, both from their full allowance of the finest grain, and in the constant health in which they are kept, by living in the natural state, and having the full enjoyment of air and exercise.

In well-fed fowls the difference will be seen, not only in the size and flesh of the fowls, but in the weight and goodness of the eggs; two of which go farther in domestic uses, than three from hens poorly fed, or half starved.

Fowls are, of all birds, the most easy to feed. Every alimentary substance agrees with them, even when buried in manure; nothing is lost to them; they are seen the whole day long incessantly busied in scratching and picking up a living.

The finest, the most imperceptible seed cannot escape their piercing eye. The fly that is most rapid in flight, cannot screen itself from the promptitude with which she darts her bill; the worm that comes to breathe at the surface of the earth, has not time to shrink from her glance; it is immediately secured by the head and drawn up.

CHAPTER III.

GENERAL VIEWS: FECUNDITY OF HENS—EXPENSES AND PROFITS.

Fecundity of hens—to promote fecundity—early hatched pullets best for breeding—will not lay while moulting—old hens cannot be depended on for eggs—to have eggs in cold weather—never allow cocks to run with hens—to promote fecundity--method adopted by the ancients—Reaumur's experiments—some hens lay more eggs than others—our experience—extraordinary products—fowls profitable.

THE question is often asked "why hens cannot be made to lay as well in the winter as in the summer?" They can, to a certain extent; but they require as a condition, that they be well provided with warm and comfortable lodging, clean apartments, plenty of food, in all its variety, consisting of grain, vegetable and animal food, pure water, and gravel, lime, and sand to roll and bask in.

A writer in the Southern Agriculturist says:—"To make hens lay in winter, they should be shut up in a warm place. Boiled potatoes, turnips, carrots, and parsnips, are cheap and good food," &c.

"The reason why hens do not lay in winter," observes a writer in the New England Farmer, "is because the earth is covered with snow so that they can find no ground or other calcareous matter to form the shells. If the bones of meat or poultry be pounded and given to them, either mixed with their food, or by itself, they will eat greedily, and lay eggs as well as in warm weather. When hens are fed on oats, they lay better than when fed on any other grain."

There seem naturally to be two seasons of the year when hens lay; early in the spring, and afterwards in summer: indicating that if fowls were left to themselves, they would, like wild birds, produce two broods in a year.

Spring-hatched birds, if kept in a warm place and fed plentifully and attended to, will generally commence laying about Christmas, or even somewhat earlier. In cold and damp this is not to be expected, and much may, in different seasons, depend on the state of the weather and the condition of the bird.

It is a well known fact, that from November to February (the very time we are in want of eggs the most), they are to many a bill of expense, without any profit. To promote fecundity and great laying in the hen, it is necessary that they be well fed on grain, boiled potatoes given to them *warm*, and occasionally animal food. In the summer, they get their supply of animal food, in the form of worms and insects, when suffered to run at large; unless their number is so great as to consume beyond the supply in their roving distance. I find it quite advantageous, in the summer, to open my gates occasionally, and give my fowls a run in the garden and field adjoining their yard, for a few hours in the day, when grasshoppers and other insects are plenty. I had two objects in view; one to benefit the fowls, the other to destroy the insects. It will be found, that the fecundity of the hen will be increased or diminished according to the supply of animal food furnished.

Hens moult and cast their feathers once every year, which generally commences in August, and lasts until late in November. It is the approach, the duration and the consequences of this period, which puts a stop to their laying. It is a critical time for all birds. All the period while it lasts, even to the time that the last feathers are replaced by new ones, till these are full grown, the wasting of the nutritive juices, prepared from the blood for the very purpose of promoting this growth, is considerable; and hence it is no wonder there should not remain enough in the body of the hen to cause her egg to grow.

Old hens cannot always be depended on for eggs in winter, they scarcely being in full feather before the last of December; and then, probably, may not begin to lay till March or April, producing not more than twenty or thirty eggs; and this is probably the cause of the disappointment of those who have supplied themselves at the markets for their stock to commence with, and get but few or no eggs. As pullets do not moult the first year, they commence laying before the older hens, and by attending to the period of hatching, eggs may be produced during the year. An early brood of chickens, therefore, by being carefully sheltered from the cold and wet, and fed once a day on boiled potatoes, *warm*, with plenty of grain, in the feeding hoppers (which will be hereafter described), and occasionally a little animal food, will begin to lay in the fall, or early in the winter.

"When," says Bosc, "it is wished to have eggs during the cold season, even in the dead of winter, it is necessary to make the fowls roost over an oven, in a stable, or to erect a stove in the poultry-house on purpose. By such methods the farmers of Auge have chickens fit for the table in the month of April, a period when they are only beginning to be hatched on the farms around Paris, although farther to the south. It would be desirable to have stoves more common in poultry-houses near cities, where luxury grudges no expense for the convenience of having fresh eggs."

A writer in the Cultivator under the signature of B., says, "I never allow cocks to run with my hens, except when I want to raise chickens." He recommends giving them fresh meat chopped fine, once a day; never allowing any eggs to remain in the nest, for nest eggs. "My hens," continues the writer, "always lay all winter, and from 75 to 100 eggs each, in succession. There being nothing to excite the animal passions, they never attempt to sit. I have for several years reduced my theory to practice, and proved its entire correctness. It must be obvious that the presence of the male is *not* necessary for the production of eggs, as they are formed whether the male be present or not. Of course such eggs will not produce chickens."

In contradiction to the foregoing, Boswell says, "To promote fecundity and great laying in the hen, nothing more is necessary than the best corn and fair water; but malted or sprouted barley has occasionally a good effect, whilst the hens are kept on solid corn, but if continued too long they are apt to scour. It must be noted, that nothing is more necessary towards success in the particular of obtaining plenty of eggs than a good attendance of cocks, especially in the cold season; and it is also especially to be observed, that a cock whilst moulting is generally useless."

"Man," says Parmentier, "who thinks of nothing but his own interest, has attempted several means of arousing hens from their torpidity, when they cease at the natural period of the year to lay, inasmuch as it seems very hard to pass through the winter without the luxury of eating new laid eggs."

The method adopted by the ancients was, rich and stimulant food, such as toasted bread soaked in ale or wine, barley half sodden, tares and millet.

M. Reaumur made several experiments with a view to the object in question. A certain class of food, and of seeds, he says, are much extolled in many places, as tending to promote the laying of eggs, but nothing has yet been determined by our choice; for in this way, the sum of the eggs laid by the hens of a poultry-yard, might be distributed in a far more equable manner, over the several months of the year; and if, as is probable, each hen can only produce a certain number of eggs, we should be glad to have a portion of them yearly produced in winter. The necessity we are under of keeping great quantities of eggs in the season when they are laid, causes an uncommon quantity to be spoiled every year, from too long keeping or want of proper caution in preserving them; and hence the importance of the question—"Whether it may not be possible to make hens lay in winter?"

With respect to fecundity, some hens will lay only one egg in three days, some every other day, others every day, and a hen was exhibited at the Fair of the American Institute, at New York, in October, 1843, that was said to have laid two eggs in a day, and Aristotle mentions a breed of Ilissian hens which laid as often as thrice a day.

According to our experience much depends on circumstances, such as climate, accommodations, feed and the attention paid to the hens, as to the number of eggs annually produced. It is asserted by Buffon, that a hen, well fed and attended, will produce upwards of 150 eggs a year, besides two broods of chickens; and a writer in the Connecticut Courant, says, "a dozen hens properly attended, will furnish a family with more than 2,000 eggs in a year, and 100 chickens," but from our experience we think this an over-estimate, especially for this cold climate. From 70 to 75 eggs per hen, a year, would be a fair estimate, for any number of fowls kept together.

We find statements from practical writers recorded in our American journals, several instances of extraordinary products of hens, which will enable us to form some judgment on the subject; but it must be borne in mind, however, that these statements have been given generally as extraordinary products.

Mr. E. A. Colman, of Chelsea, Mass., obtained from eight hens, from July 7th to August 29th, seven weeks, 293 eggs.—*N. E. Farmer.*

The editor of the Newburyport Herald gives an account of an experiment made with ten hens which produced 1,116 eggs, besides 15 chickens, within the year. The avails of 41 dozen eggs sold, paid all the expense of keeping.

It is stated in the Farmer's Journal that from 150 hens, 1,900 eggs were obtained in the month of January; and that five pullets produced 300 eggs from the middle of October, to the middle of April, which is the coldest part of the year.

"Mr. E. Tucker, of Milton," says the editor of the Massachusetts Ploughman, "tells us that he obtained 600 dozen (7,200) eggs in one year from 83 hens; this was his highest number of fowls; he sometimes had less; that for 562 dozen (6,744) he took $100 within one cent. The whole amount of his cash expenditures was $56 43, leaving him a balance of $43 57."

Mr. Westfall, of Rhinebeck. in the Cultivator,

says, "From 45 hens, I have the past year (1840) raised more than 150 chickens, although I had rather poor success in hatching the eggs. I have sold eggs to the amount of $21 29: have now over 300 on hand, and the year since the receipt of the first egg last spring will not be up till the last of this month (February), and we are now getting from 20 to 25 eggs per day from about 80 hens."

A writer in the Cultivator of 1842, says, "This year (1842) I have about 40 hens, mostly pullets, and three cocks. They commenced laying in the latter part of January, and, up to the last of April, about 90 days, have given me about 120 dozen (1,440) eggs."

A correspondent in the Cultivator who writes more in detail, says, "that when his fowls commenced laying he had 37 hens and three cocks, and when they ceased laying he had 26, the average being 32. In about 300 days, between January and November, they yielded 3,298 eggs." He raised no chickens.

The following remarkable instance of fecundity is furnished by Mr. Morent, in the Cultivator. He had three pullets of the Poland or top-knot variety, which were hatched in June. December the 15th of the same year, they began to lay, and from that time to December following, laid 524 eggs. Cost of keeping not exceeding $3 71. They were fed on barley, rice and peas.

A correspondent of the Massachusetts Ploughman gives a most extraordinary account of the sales from the produce of one hen, amounting to the sum of eighteen dollars.

The following singular case is related in an English publication. "Mr. James Drinkwater, of Harpenbery, has a hen two years old; it has not a white feather on it, but is as black as jet. For upwards of eighteen months it has laid an egg every other day, and has never been known to change its feathers."

A friend living on Staten Island, and who has been pretty successful in the management of poultry, informed me that from 55 hens and 7 ducks, he obtained in the months of January (1842), 182 eggs—in February, 324—March, 792—April, 878—May, 915—June, 746—July, 534—August, 650—September, 346—October, 68—November, 5—December, 69—making in all for the year 5,509. Allowing the seven ducks to have laid 70 eggs each, would leave 5,019, which divided by 55 gives an average of 91 eggs to each hen. These hens were fed from six to eight quarts of cracked corn per day, and occasionally a few boiled potatoes. Averaging the feed at 7 quarts per day we have within a fraction of 80 bushels of corn, which at 50 cents per bushel amounts to $40, and allowing the eggs to be worth $1 per 100 we have $55 09, from which deduct $40 for food and we have a profit of only $15 09 besides 60 chickens, which, at 15 cents each, would swell the profits up to $24 09. He gives the preference, for eggs, to the silver top-knots and French hens.

Another friend who resides in a neighboring city and keeps between 30 and 40 hens, informed m that, in 1841, he obtained eggs from his hens through out the year; that is, there was not a single day in which he did not obtain some. This he accounted for by having very early chickens, as when the old hens ceased laying to moult, the young pullets commenced. In 1842 he kept between 25 and 30 hens, and obtained 2,832 eggs. This, it will be seen, gives a fraction over 94 eggs to each hen, which is nearly double the number we obtained from our hens. His yard is completely protected by high fences and buildings on the north and west, and receiving the full influence of the sun from the south. He has low sheds around the yard to protect them from storms in the day and a warm room in the loft of his wood-house, which is lathed and plastered, to protect them from vermin, and keep them warm in cold weather.

In 1840 my hens commenced laying on the 7th of February, and between that period and the 15th of August, when they commenced moulting, we obtained 2,655 eggs from 60 hens; when the year previous, from 100 hens, which were suffered to run at large, we did not get but few over 1,000. In 1841 they commenced laying the 8th of January, and continued to lay until the 27th of September, when they ceased entirely, but commenced again on the 13th of Octo ber, and continued to lay until the 18th of November, when they ceased, and commenced again on the 1st of December; and up to the first of January, they produced over 4,000 eggs. In 1842 we had 71 hens, which produced within the year 3,509 eggs. In 1843 we kept 69 hens, from which we obtained 3,978 eggs.

In order to ascertain by demonstration, and to satisfy ourself, whether the keeping of fowls were profitable or not, we commenced in 1842 keeping debit and credit account with the poultry-yard. We had 71 hens, 12 cocks, 2 ducks, 2 drakes, 3 turkeys, 1 turkey-cock, 5 geese, and 2 ganders—in all 98 head.

They consumed within the year as follows:

91 bushels	Wheat Screenings,	at 21 cts.	$19	11
6 "	Rye,	5s	3	75
11 "	Millet,	5s	6	62
2 "	Indian corn,	5s	1	25
3 "	Barley,	4s	1	50
2 "	Indian Meal,	8s	2	00
10 "	Small Potatoes,	1s	1	25
			$35	48

We obtained 3509 eggs valued at	$35	09
Sold fowls for	2	00
" 3 Geese,	2	50
" 6 " consumed in family,	2	25
" 5 Turkeys do	1	87½
" 30 Fowls do	5	62½
" 60 Ducks' Eggs do		83
" 54 Geese do do	1	62½
" 8 lbs. Geese Feathers at 5s	5	00
	56	79
From which deduct	35	48
Nett profit,	$21	31

Now from the foregoing it would seem that the profits are very small, but it must be recollected that the sale prices are very low, and that we had the misfortune to lose many of our chickens by the hawks, and the greater part of our goslings by confining them in a yard, when they should have had the run of a pasture, which would have saved considerable food, and probably the lives of the goslings, and would have made some difference on the credit side. We lost many of our turkeys in the same way. The experience of this year taught us that it will not answer to keep or confine goslings and turkeys after they are half grown.

In 1843, the care of the poultry was entrusted to my son, a lad 15 years of age, and the following is his account rendered on the 1st of January, 1844:

Poultry Yard, *Dr.*

To 69	Hens on hand, valued			$25 87½
15	Cocks,			7 50
3	Turkeys,			1 87½
7	Geese,			7 00
1	Fancy Duck,			1 00
1	Guinea Fowl,			25
71	bushels	Wheat Screenings,	at 15¾ cts.	11 25
4	"	Millet,	50 cts.	2 00
14½	"	Corn,	42 cts.	6 09
30¼	"	Oats,	24 cts.	7 26
8	"	Potatoes, small	2s	2 00
32	Fowls purchased,			15 09
3	Turkeys, do			1 13
				$88 32

	Poultry-yard	*Cr*
By 3978	Hens Eggs, at 1 ct	$39 78
175	Turkeys, Geese and Ducks' Eggs,	2 56
41	Fowls sold for	46 31
30	do. consumed in family,	7 06
11	Died. 5 presented to friends.	
5	Geese sold,	7 00
2	do. consumed,	2 00
2	Turkeys, do.	1 00
1	Died.	
32	bushels manure, sold for	6 00
54	Hens on hand,	20 25
18	Cocks,	9 00
6	Geese,	6 00
1	Duck,	1 00
10	Turkeys,	5 00
2	Guinea Fowls,	50
		153 46
	Deduct	88 32
	Nett profit	$65 14

It is stated in the report of a committee on poultry of the Wayne County (N. Y.) Agricultural Society, that David Cushing keeps 25 hens, and feeds them with oats, corn-meal, broom-corn seed and refuse meat, and supplied with ashes, pounded shells, &c. They were confined to a warm but dry room in winter. His account is as follows:

Poultry	*Dr.*
To investment of stock and fixtures,	$50 00
Interest,	3 50
Feed, 25 bushels of Oats, at 20c. (large estimate)	5 00
Attendance,	5 00
	$63 50

Poultry Establishment		*Cr.*
By 75 Doz. Eggs, sold early, 12 cts.	$9 38	
200 Chickens, 10 cts.	20 00	
Stock and fixtures on hand,	50 00	$79 38
Leaving a nett balance of		$15 88

on an investment of $50, or an interest of more than 25 per cent. on the capital employed.

J. J. Thomas, of Macedon, states that carefully conducted experiments have led to the conclusion, that "$1\frac{1}{2}$ bushel of oats will be an adequate yearly supply for each adult" hen, and that by allowing the fowls "an hour's run for exercise before retiring for the night, high health and productiveness will be secured."—*Cult.*

A writer in the New Genesee Farmer says, "I had 60 hens; they have brought me 500 dozen eggs the past year. One hen stole her nest and raised five chickens. Average price of eggs 17 cents.

	Cr.
500 doz. sold at 17 cts.	$85 00
5 Chickens, at 2s	1 25
	86 25
Expense of grain,	55 00
Nett profit,	$31 25

CHAPTER IV.

FOOD.

Food—amount of grain consumed—experiments with boiled grain—kinds of grain preferred—green food—butcher's meat, fowl and fish—brewer's grains good for fowls—Mr. Hunt's method of feeding to promote laying—good effects of extra attention to fowls in cold weather—good effects of fowls running in the garden—fattening—should prefer moderately fat—general method of fattening—best method—Bradley's plan— Wakefield's—Wingate's—peculiar advantages—quality of food should be attended to—should be varied—cramming fowls disgusting—description or coops or hatches for fattening—flavor of meat obtained by feeding with spices and aromatics—wild fowl affected by eating aromatic plants—turkey's flesh tinctured by eating onion tops.

It is customary to throw to the fowls in a poultry-yard, once or twice a day, a quantity of grain, generally corn, and somewhat less than that which they would consume if they had an abundance. Fowls, however, are more easily satisfied than might be supposed from the greedy voracity which they exhibit when they are fed. It is well known that, as a general rule, large animals consume more than small ones. There is as much difference in the quantity of food consumed by individual fowls, as there is in animals.

I was curious to ascertain the quantity of each sort of grain which a given number of fowls when abundantly supplied would consume; and for that purpose, I confined one cock and seven hens of the Poland variety. The first feed I gave them was one peck of Indian corn, which they consumed in eleven days. I then fed them one peck of oats, which they ate in six days. The next feed was the same quantity of barley, which lasted them seven days. The like quantity of wheat, they consumed in ten days. The same quantity of millet lasted them eight days, and the like quantity of wheat-screenings they devoured in seven days. During this trial they had no other food, except a few boiled potatoes.

It has been ascertained by actual experiment, that, in the months of January and February, a common sized fowl will consume, when at command, of corn, wheat, rye, or barley, about one gill per day. M. Reaumur instituted a series of experiments to ascertain the quantity of each sort of grain which a fowl would consume, when abundantly supplied therewith from morning till night; and in the course of his experiments, he discovered many interesting particulars, of importance to be known to all those who keep poultry for profit. He found that individual fowls vary very much in the quantity of food which they consume—there being little and great eaters among them, most commonly indicated by the size of the body; that two bantams might be kept on the same amount of food as one of the largest breed. Even among fowls of the same size and kind, there are individuals which require more food than others a circumstance that can be only ascertained by trial.

For the purpose of accurately ascertaining the quantities of food consumed, M. Reaumur confined fowls separately under basket coops; and others in hutches enclosed with wooden gratings, where they had more convenience, even so much as to lay eggs there in the same way as if they had been at liberty. To the hens in each hutch, he put a cock, in order that nothing might be wanting to the completeness of his experiments. In some hutches he placed as many as seven hens, and in others as few as two. For several days together, he gave both to the fowls in the basket coops, and to those that lived in company in the hutches, the same quantity of grain, measured so as to be more than would fill their crops; and care was taken that the box into which the grain was put for them should never be empty. This box was longer than broad, with a bottom, and a piece of a board on each side, projecting about five or six inches, so fixed as to prevent the chance of its being upset by the fowls hopping upon it, while the sides were sufficiently high not to allow them to scrape the grain out of the box,—precautions indispensable to the accuracy of the experiments, as in this way every grain of corn could be accounted for. Gravel was also spread upon the bottom of the hutches and coops, and some was placed in a separate vessel, as being judged indispensable to promote digestion.

Nearly the same measure of grain was found sufficient for a fowl every day, whether it consisted of oats, buckwheat, or barley; and hence, whichever of these three is cheapest at any time, may be given, without regard to other considerations. Variations in the appetite of the fowls may, perhaps, be occasioned by difference of seasons, and they may require rather more at one period than another; but it was ascertained that in the months of January and February, a common barn-door fowl, that has always, from morning till night, grain of one of these three sorts at command, will eat of it daily about a fourth part of a pint measure. This is even rather more than an ordinary sized fowl will eat, for when a quart was given to a very large cock and Spanish hen, and the same to two hens of a middle size, and to three of the ordinary size, it was not all eaten. Some very voracious fowls of the largest size, however, will consume daily about the third of a pint measure.

As wheat is the most nutritive grain for human food, with the exception perhaps of rice, it might be supposed that it is also the best for fowls; and as they will eat wheat greedily, we might thence be induced to conclude that they would eat more of it than of barley or oats. Yet when fowls have as much wheat as they can consume, they will eat about a fourth part less than of oats, barley, or buckwheat; the largest quantity of wheat eaten by a fowl in one day being about three-sixteenths of a pint; nevertheless, the difference of bulk is compensated by the difference in weight, for these three-sixteenths of wheat will weigh more than six-tenths of oats.

The difference in weight in different sorts of grain is not in every instance the true reason why a fowl is satisfied with a larger or smaller measure of one sort than another; for though rye weighs rather less than wheat, a fowl will be satisfied with a much smaller measure of this—even, in most cases, so little as one half. The seven hens and the large cock just mentioned, consumed daily a pint and a half measure of wheat, while of rye they only consumed three quarters of a pint measure, and hence the average consumption of the rye by each, was to their consumption of wheat in the proportion of one to two.

Indian corn was found to rank intermediate between rye and wheat. When corn was exclusively given, the greatest eaters only consumed the first day about one-eighth of a pint measure, but by usage they came to relish it more; and the cock and seven hens, which were rather above the average rate of eaters, consumed daily one pint and a quarter of corn. Accordingly, five-fourths of corn to them were equivalent to six-fourths of wheat, and to three-fourths of rye.

The consumption of each sort of grain daily, by a common barn-door fowl, will be rather too great, if we take the data furnished by what was taken by the cock and seven hens, as some of these were of very large size, and great eaters; though it is more convenient for the practical purpose of estimating expense, to be above rather than below the average; what is spent less than what one is willing to spend, becomes, in one sense, clear profit. We may therefore safely estimate, that a barn-door fowl of the common size, having as much as she can eat during the day, will consume

	Pint measure
Of oats, barley, or buckwheat	8-32
Of wheat	6-32
Of corn	5-32
Of rye	3-32

Although, from the experiments already detailed, as made with wheat and rye, it appears that the average consumption is not always in proportion to the specified weight of the corn; yet it is of importance to know the relative weights of each grain in all such experiments. M. Reaumur, in order to ascertain the difference of weight of each grain in different circumstances, carefully weighed a pint measure of each; when he found the weights to be the following:

	oz.	dr.	gr.
Wheat	19	1	52
Rye	18	4	12
Corn	17	5	48
Buckwheat	16	7	12
Barley	14	0	48
Oats	10	3	12

After weighing, each of the sorts of grain was put separately into a paper bag, and laid in a low and very damp room for nearly two months, when they were again weighed. The measure, which had been struck in the first instance, was now found to have the corn considerably above the brim, owing to the increase of bulk from moisture. The weights were the following:

	oz.	dr.	gr.
Wheat	18	1	54
Rye	18	1	18
Corn	16	3	18
Buckwheat	15	7	34
Barley	12	6	6
Oats	9	15	12

These tables show that buckwheat is considerably heavier than barley and oats; yet a fowl will require as much by measure of buckwheat daily to satisfy her as of any other two sorts of grain. The cause of this apparent anomaly may, perhaps, lie in the peculiarity of taste, one sort of grain being more pleasing to their palate than another, and inducing them to eat gluttonously more than might suffice them, in the same way as dainties will induce ourselves to eat more than nature requires. In order to determine, if possible, the case in question, M. Reaumur had a partition made in the feeding boxes, dividing them into two equal compartments, into one of which he put a measure of wheat, and into the other a measure of rye. Three hens and one cock were fed upon them, and did not show any preference to either of the sorts of grain, as there remained after their first morning's repast, about an equal quantity of wheat and of rye; while in the course of the day they finished what they had left, emptying the two compartments at almost the same time.

Experiments of this kind, which ought to be repeated with all the sorts of food given to poultry, are important for pointing out what sort of food is most to their liking; a matter of great moment in feeding, as it is a rule, with but few exceptions, that what is eaten with most relish agrees best, and is most easily digested. With a view to this point, M. Reaumur extended his experiments to other sorts of grain, by not only offering the fowls each sort in different compartments, but also by putting a mixture of grains into the same box, and mixed them. A cock and a hen, to which this mixture was given, exhibited a greater liking to the wheat than to the oats, for there remained at night in the box a portion of the oats, about a fourth or a sixth part of the whole quantity, but there was not left a single grain of wheat. From a subsequent experiment, however, it appeared that it would have been too hasty to draw a general conclusion from the taste of this cock and hen. On the same day, there was given to a hen kept alone under a basket coop, a measure of wheat in one compartment, and a measure of oats in the other compartment of the feeding box. In the course of the day, this hen ate the whole of the oats to a grain, and left almost half of the wheat. A measure of rye was given her next day, and she preferred eating that rather than the portion of wheat she had left the day before. To the same hen was given a measure of wheat and oats mixed in equal quantities. At first she was seen to eat both sorts of grain, but as she continued to feed the oats began to disappear, and the wheat predominate, till at length every grain of oats was consumed, and about one-sixth of the wheat was left untouched.

At another time, M. Reaumur gave to a cock and a hen, kept in the same hutch, three different sorts of grain, namely, corn, oats, and buckwheat, each put separately into three different compartments of the feeding box. The cock came first to the corn, and after eyeing it for a time, he finally refused to touch it, but turned to the buckwheat, of which he picked up a few grains, and then went to the oats. He continued for some minutes to taste a little of the one, and a little of the other. On the other hand, the hen no sooner saw the corn than she pounced upon it voraciously, and never offered to quit it for either of the others. As the hutch confined their motions, he tried the effect of turning the box, so as to bring the corn opposite the cock, and the others more within reach of the hen; but the cock even then never offered to touch the corn, but went round to the other two sorts, while the hen also changed sides, and recommenced eating the corn with avidity. Next day, the same cock and hen were served with the three sorts of grain as before; but now the hen seemed as little disposed as the cock to try the corn, but fixed upon the buckwheat, which between them was wholly consumed What had been left of the oats was next eaten, and in the course of the day the corn was also consumed. It is necessary to remark, that the corn which is yellow-colored is not so good as that which is reddish, or rather reddish brown.

These experiments were varied and extended by giving to the seven hens, already mentioned, an equal quantity of each of the six sorts of grain usually given to poultry, each part separately into the compartments of a common feeding box, the whole quantity given being calculated to serve two days. The first day, the whole of the buckwheat was despatched; on the second morning there was no barley left; the wheat and the corn were entirely consumed nearly at the same time, on the second afternoon; while a little rye and oats were left for the third day.

It would be superfluous to detail, with similar minuteness, the numerous experiments of the same kind, which were varied and combined in every possible way that M. Reaumur could devise, as he found it was by no means as easy as might have been previously supposed, to determine the sorts of grain which fowls prefer. At first, it appeared that there were some hens which ate more readily any particular sort they had been most used to, than sorts entirely new to them. On the contrary, it also appeared that the appetite of others was excited by any new sort. In a word, similar singularities of taste appear amongst fowls, with respect to particular sorts of food, as amongst ourselves. It is very certain, at least, it is not because one sort of grain is larger, heavier, or plumper than another, that they eat more or less of it, or prefer it to others, but it is probable that the grain of which they consume the least furnishes the greater portion of chyle—the bland creamy fluid prepared by digestion to mix with the blood, for promoting the growth and repairing the waste of the body—in a word, that which supplies the most nourishment.

Further experiments proved that the sorts of food most easily digested by fowls, are those of which they eat the greatest quantity. Of the six sorts of grain already mentioned, they evidently became soonest tired of, and least partial to rye.

EXPERIMENTS WITH BOILED GRAINS.

It is the custom of poultry-keepers in France, to cook the grain given to fowls which they intend to fatten, boiling it in water till it is soft enough to be easily bruised between the fingers, the boiling causing it to swell till the farina splits the enveloping membrane, and this they term bursting. Although it is the popular opinion that burst grain is better than when it is dry, for fattening poultry, this opinion has probably not been established on accurate experiments. Be this as it may, it is of no less importance to ascertain whether there is any difference of expense in feeding poultry on dry or on burst grain, that is, whether, under similar circumstances, fowls eat more or less of the one or of the other.

"In order to ascertain this," says Dickson, "M. Reaumur ordered four measures of each of the six common sorts of grain to be boiled till they were well burst, and he found that the increase of bulk in each sort was the following:

	Measures
Four pint measures of oats, after being boiled to bursting, filled	7
Four pint measures of barley, after being boiled to bursting, filled	10
Four pint measures of buckwheat, after being boiled to bursting, filled	14
Four pint measures of Indian corn, after being boiled to bursting, filled	15
Four pint measures of wheat, after being boiled to bursting, filled a little more than	10
Four pint measures of rye, after being boiled to bursting, filled nearly	15

Rice swelled considerably more by boiling than any of these six sorts, but it is rarely given to poultry, except for fattening, under the notion that it tends to whiten the flesh.

For the purpose of ascertaining whether the boiling altered the preference of fowls for any of the particular sorts, experiments, varied in every possible way, similar to those already detailed, were made by M. Reaumur. The fowls were furnished with two, three, four, five, and six different sorts, sometimes all the compartments of the feeding box being filled with burst grain, each different from the other, and sometimes each sort of grain filled twc of the compartments, one of them having nothing but boiled, and another nothing but dry grain. All that could be collected from these repeated experiments, was, that the greater number of fowls prefer boiled grain to raw, though there are many of them which show a preference to the dry grain on certain days, and no permanency could be discovered in the preference shown for any sort of burst grain. Some fowls, for example, which one day preferred boiled wheat, would on other days make choice of buckwheat, or corn, oats, or barley, and sometimes, though more seldom, even of rye but rye, either boiled or raw, is the least favorite sort of grain. It follows, as an important conclusion from such experiments, that we may make choice of the sorts of grain which happen to be cheapest, without much, if any disadvantage; always excepting rye, when other sorts are to be had on reasonable terms.

Other experiments were required to show whether there is any economy, or the contrary, in feeding poultry with boiled grain, and this was readily ascertained from knowing first, how much dry grain sufficed one or more fowls, and then boiling the same quantity, and trying how much of that would in like manner be sufficient. The experiments made with the different sorts of grain, were as follows:

Rye, although so very considerably increased in bulk by boiling, so far from being more sufficing, became less so, as fowls will eat rather more of it when it is boiled than when it is dry. The seven hens and the cock, so often mentioned, consuming only three-fourths of a pint measure of dry, it would cost one-twentieth more to feed fowls with boiled than with dry rye, four-fifths being one-twentieth more than three-fourths.

Oats, although increased in bulk by boiling nearly one half, are not, any more than rye, rendered more sufficing; for the fowls which in two days would have eaten four pint measures of dry oats, consumed in the same time seven pint measures of the boiled grain,—consequently it is no saving to boil the oats. —*Dickson.*

Mowbray says oats are apt to produce scour, and chickens become tired of them; but are recommended by many for promoting laying, and in Kent, Sussex, and Surrey, for fattening.

Buckwheat is increased in bulk by boiling still more than oats, as four pint measures, when well boiled, swell to fourteen; yet is there small benefit obtained by boiling buckwheat; for the fowls consume the fourteen pint measures of the boiled grain, nearly in the same time which four pints of the dry would have sufficed them. Mowbray says it is an unsubstantial food.

Indian Corn is, on the other hand, more profitable when boiled than when given raw; for the fowls which would have got through a pint and a quarter of the dry corn, consumed only three pint measures of the boiled grain, which are not equivalent to one of the dry. It was for two days only that they were able to eat in a single day three pint measures of the boiled corn; for after that, they either lost their appetite, or came to dislike it, as they could not get through more than two pint measures of the boiled corn. Even calculating that they were to consume three pint measures a day of the boiled grain, there would be a saving of more than one-fifth, and if they were satisfied with two pint measures, the profit would be much more considerable; for this would not be equivalent to two-thirds of a pint measure of the dry grain. The saving in this case would be one-third and one-fifth, that is, eight-fifteenths, or more than one half.

Barley is also much more economical when given boiled than dry; for fowls which would have eaten two pint measures of dry barley a day, ate but three pint measures of boiled grain. Therefore, as ten pint measures of boiled barley are produced from four pint measures of dry, three pints of the boiled are equivalent to no more than six-fifths of a pint of the dry; consequently, the expense in dry barley is to that of boiled as ten-fifths ($\frac{10}{5}$) to six-fifths, that is, as ten to six, or as five to three, showing a saving of two-fifths by giving boiled instead of dry barley.

Wheat is shown by the preceding table to increase in bulk by boiling about the same as barley; but experiments prove that the saving to be obtained by feeding fowls with boiled wheat, is not nearly so much as might thence have been anticipated; for the same fowls which consumed three pint measures of boiled barley in one day, ate three pint measures of boiled wheat. Three pints of boiled wheat, however, are not equivalent to two pints of dry wheat, as in the case of the barley, but only one pint and a half of dry wheat, which was found to be the quantity consumed in one day by the same fowls. Now, as a pint of boiled wheat is equivalent to no more than two-fifths of a pint of the dry grain, the three pints consumed a day are equivalent only to six-fifths of dry wheat. Consequently, the proportion of what they consumed of dry corn was, to what they ate of boiled, as fifteen-tenths to twelve-tenths, or as five to four; hence there is a saving of one-fifth by feeding with boiled wheat, as there is of two-fifths by feeding with boiled barley.—*Dickson.*

These interesting experiments prove most clearly that in every case, where the price of corn, barley, or wheat, renders it eligible to feed poultry therewith, there is considerable economy in never giving the grain dry, but well boiled; and there is no saving by boiling oats, buckwheat or rye.

Rice.—Boiled rice might be supposed to be a very nourishing food for poultry, though it is too expensive for daily feeding, and they are at first very fond of it; but their liking for rice does not continue, and in a week or so they come to dislike it. One reason may be that it is too clogging; and were it mixed with some less nourishing substance, such as bran, the fowls would continue to relish it as they do barley.

Millet.—Fowls prefer raw millet to that which has been boiled, though it would evidently be a saving in other respects to boil it, as boiling increases its bulk above one half. We have found millet excellent food for young chickens.

Potatoes.—As potatoes contain a great proportion

of nutriment comparatively to their bulk and price, they constitute one of the most economical articles upon which poultry can be fed. The poultry-keepers in England consider potatoes excellent for promoting laying in fowls; while M. Parmentier advises that they should only be given for the purpose of fattening, since he thinks they will render the fowls so fat as to hinder them from laying.

Potatoes are, according to our experience, a cheap, wholesome, and nutritious food for fowls, though it would require experiments similar to those of M. Reaumur, already detailed, with respect to grain, to ascertain the quantity which each fowl would consume when potatoes are supplied in over-abundance. If fed alone, without grain, they are very apt to make them scour. And we have found it indispensable not only to feed them in a boiled state, but *hot;* not too hot, however, as they are so stupid as to burn their mouths if permitted. It is likewise necessary to break the potatoes a little, for they will not unfrequently leave a potato when thrown down unbroken; taking it, I presume, for a stone, since the moment the skin is broken, and the white of the interior is brought into view, they will pounce upon it greedily.

Fowls are not fond of raw potatoes, beets, carrots, or parsnips, though they will sometimes eat them, when cut into very small pieces. Boiled vegetables mashed up with bran or meal, are excellent food for poultry, and answer well for their evening meal, when grain has been given them in the morning.

Green Food.—From seeing fowls, when at liberty devour plants and leaves, it is generally supposed that they will eat anything that is green; but such is not the case, as I have found by experiment. Among the plants which they reject, are the leaves of strawberries, celery, parsnips, carrots, potatoes, and convolvulus. They are most partial to the leaves of lettuce, endive, cabbage, spinach, raddish, turnip, and chick-weed. They also eat grass, and M. Reaumur says, "that if hens have green plots to go a grazing in, from morning till night, which they are naturally inclined to do, and which they will be naturally compelled to do if they are sparingly fed with grain, the expense of keeping them will not be half what it would be as when they are furnished with as much boiled barley as they choose to eat."

Poultry, however, are not the better for being fed entirely on raw greens, as it is very apt to relax and scour them, and cabbage and spinach are still more relaxing to them when boiled than raw. M. Parmentier recommends, and this accords with our own experience, giving them all the refuse of the kitchen, such as bits of spoiled fruit, parings of apples, and the like; but I have found my fowls are not fond of the latter.

The left pieces and crumbs of bread, pie-crust, fragments of pudding and dumplings, all fowls are very fond of. There can be little doubt but that biscuit-dust from ship-stores, which consists of biscuits mouldered into meal, mixed with fragments still unbroken, would be an excellent food for poultry, if soaked in boiling water, and given them hot. It can sometimes be had in large sea-ports, in considerable quantity, and at a very reasonable price. It will be no detriment to this material, though it be full of weevils and their grubs, of which fowls are fonder than of the biscuit itself.

Butcher's-meat, Fowl, and Fish.—A fowl appears to be delighted when, after having scratched up the ground, she discovers an earth-worm, on which she does not fail to pounce with avidity; and from the ravenous voracity with which they pounce upon any scrap of meat they meet with, we might suppose that they are more carnivorous than granivorous. This, however, is only observed from the meat being an occasional tit-bit. Were they fed entirely on meat, without any grain, for some time, they would manifest the same voracity for the latter. But it is well to take advantage of this omnivorous propensity to make use of every scrap of meat and offal which would otherwise be lost, as such must always assist in saving the quantity of corn which they would otherwise require. Fish is no less wholesome to them than flesh, and they are as fond of it salted as fresh.

It seems to make but little difference with them, whether any sort of animal food is raw or boiled, though perhaps what is raw is more highly relished; at least they are fond of blood, which they will sip up from the ground where it has been shed, till not a drop remains.

Pieces of suet or fat, they like better than any other sort of animal food; but this, if supplied in any quantity, will soon render them too fat for continuing to lay.

There is no sort of insect, perhaps, which fowls will not eat. They are exceedingly fond of flies, beetles, grasshoppers, crickets, and every sort of grub and maggot. We found it quite advantageous in the summer to open our gates occasionally, and give the fowls a run in the garden an hour or so, in the afternoon, when insects and grasshoppers are plenty.

Mr. Stimson, of Galway, a few years since connected the business of rearing poultry with the useful purpose of protecting his garden from the depredations of the numerous tribes of insects which so frequently render abortive the best exertions of the gardener. His method is simply this: a sufficient number of coops are constructed, and are placed in different parts of the garden, and the hens with their broods are put into these coops; the chickens, finding no restraint on their freedom, roam over the garden, and devour every fly, bug, or insect, which falls in their way. There is one objection, however, to this, which we found by experience, and that is, if left in the garden too long, they become so attached that it is difficult to keep them out when grown up. We would therefore recommend removing them to the poultry-yard as soon as they get in feather.

A writer in the New England Farmer says, "I keep my hens warm under cover during the winter, and feed them on 'brewer's grains,' placed in an opei box or tub, that they may eat when they please, occasionally giving them oats, corn, and oyster-shells, pounded fine, and plenty of water. By keeping them well fed and warm, they begin laying earlier in the season. I prefer spring chickens, as they lay earlier than old hens—and the old hens sit more steady, and make the best of mothers."

Mr. Hunt observes, "I have often been asked, what is the best food to make hens lay? I have made several repeated experiments to decide this question. The result is, give your hens and rooster (who, by the way, requires as much, nay, more attention than the hens), water, gravel, and animal food, and they will lay as well on one kind of food as another. Boiled potatoes, corn, wheat, rye, barley, buckwheat and anything that they eat will do. Boiled food is cheapest and best for hens, especially if kept up all the year, as they should be.

"I have followed the above rule ever since I have owned chickens. We have always had more eggs than required for use; and our chickens have never had any epidemic among them. With the exception of the moulting season, that is, when they shed their feathers, with judicious management, hens will lay for two hundred and sixty days in the year."

The Editor of the Farmer's Journal remarks, "We will name an instance of the good effects of extra attention to fowls in cold weather. We had a lot which were supplied with grain, water and gravel, in the cold season. They did not lay till the latter part of February,—they were old hens. The next winter, in addition to grain, we gave them warm food of potatoes, meal, &c., a fresh lot of gravel every week, and pounded bones and oyster-shells, and care was taken to keep the hen-house clean. In January, the second winter, the same hens laid abundantly. The eggs were worth three times as much as the food consumed."

FATTENING.

The substance termed *fat*, in any sort of animal, does not form a necessary part of the body, but is only the superabundant nourishment, which is not wanted for repairing the wear and waste of the bones, muscles and other solids. Fat, accordingly, in a living animal, has no more sensibility than the hair or the nails, and may, therefore, be cut into without giving the least pain, provided no other part besides the fat be injured. It is stowed up in membraneous cells in various parts of the body, in readiness to be taken up by the absorbent vessels, and turned to use, whenever the supply of nourishment, prepared by the stomach, and other organs of digestion, becomes deficient.—*Dickson.*

In order to make an animal fat, it is requisite that nourishment be supplied more plentifully than is wanted for ordinary nutrition ; or, at least, more plentifully than the absorbent vessels can take up and dispose of; consequently, if the activity of the absorbent vessels is diminished by any cause below its natural standard, fat must necessarily accumulate, till that activity be restored.

One principle, upon which the absorbent vessels are rendered less active, consists in avoiding all stimulants and all exercise, even the stimulus of light, and the motions of the body, with everything that might produce irritation or give uneasiness. Another principle is, to load the stomach with the most nutritious food, provided it be bland, and without stimulus.

Fowls, as well as other animals, ought to be preferred when they are moderately fat, rather than when they have been too high kept.

These general remarks will enable us to appreciate the various modes for fattening fowls, which have been practised and recommended, which we shall now fully detail.

The points for consideration on this branch of the subject are, the local *conveniences;* the modes, common or extraordinary; the variety and quality of the *food;* and the length of *time* necessary for completion of the object.

The well-known common methods are to give fowls the run of the farm-yard, where they thrive upon the offals of the stable and other refuse, with perhaps some small regular feeds; but at threshing time they become fat, and are probably the most delicate and high-flavored of all others, both from their full allowance of the finest grain, and the constant health in which they are kept, by living in their natural state, and having the full enjoyment of air and exercise; or they are confined during a certain number of weeks in coops. It is a common practice with some to coop their fowls for a week or two, under the notion of improving them for the table, and increasing their fat; a practice which, however, seldom succeeds, since the fowls generally pine for their loss of liberty, and slighting their food, lose instead of gaining additional flesh. Such a period, in fact, is too short for them to become accustomed to confinement.—*Mowbray.*

"To fatten fowls," says Bradley, "the best way, and quickly, is to put them in coops as usual, and feed them with barley-meal; but, in particular, to put a small quantity of brick-dust in their water, which they should never be without. This last will give them an appetite for their food, and fattens them very soon." He thinks the brick-dust acts as gravel is so universally supposed to do, in grinding or bruising the food in the gizzard.

"In his extensive establishment near Liverpool," says Dickson, "Mr. Wakefield fattens with steamed or roasted potatoes given *warm*, which is indispensable, three or four times a day. The fowls were taken in good condition from the yard, confined in dry, well-ventilated coops, and covered in, so as to prevent the entrance of too much light. This method was attended with the greatest success."

P. Wingate of Maine, who has been very successful in his practice, says he shuts them up where they can get no gravel. He keeps corn by them all the time; gives them meal mixed up with water in the form of dough, and instead of water gives them skimmed milk. They become fat in ten days. If they are kept longer than ten days they must have gravel, or they will fall away.

All practical and practicable plans have their peculiar advantages; among others that of leaving poultry to forage and shift for themselves; but where a steady and regular profit is required from them, the best method, according to my experience, whether foı domestic use or sale, is constant and high keep from the beginning, whence they will not only be always ready for the table with very little extra attention, but their flesh will be superior in juiciness and rich flavor to those which are fattened from a low and emaciated state. Fed in this mode, the spring pullets are particularly fine, at the same time most nourishing and restorative food.

The quality of the food is a principal thing to be attended to in fattening poultry; for if any of the ingredients be otherwise than sweet, the flesh of the fowls will be deteriorated. Many imagine that, if fed with anything they will eat, they will answer the purpose of the feeder; but this is a mistake, for every one knows that has fed fish to fowls, their flesh becomes so highly tainted with them as to render them unfit for the table. We have been credibly informed that hens fed on onions imparted the flavor to the eggs so strongly that they could not be used.

In fattening fowls we prefer to vary their food daily, for we do not believe any one article is of itself best adapted to fatten poultry. They should have corn (cracked), wheat, barley, buckwheat, millet, peas, if boiled so much the better; ground oats, barley or meal mixed with molasses, coarse sugar, mutton suet, sheep's liver chopped fine; boiled potatoes with Indian-meal mixed, and milk-curd.

Instead of giving ordinary light grain to my fattening fowls, I have always found it most advantageous to allow them the heaviest and best. This high feeding shows itself not only in the size and flesh of the fowls, but in the size, weight, and substantial goodness of their eggs, which, in these valuable particulars, will prove far superior to the eggs of fowls scantily fed or on soft and washy substances—two eggs of the former going farther in domestic use than three of the latter. The water, also, given to fattening fowls should be often renewed, fresh and clean.

The disgusting method of cramming fowls, as adopted in France and England, has not as yet, I believe, been introduced here, and for the honor of our country, I hope it never will be adopted.

For the benefit of those who are desirous of trying coops or hutches for fattening fowls, we will give a description of one we saw at the mansion of Mr. Geo. Law, of Baltimore. It is a box raised about two feet from the floor, standing on small smooth legs, so that neither cats nor mice can climb up, and divided into compartments to hold one fowl each. The front is a door that slides up and down, with two slits about two inches wide running up and down, to allow the fowls to feed out of a narrow trough which contains water and food. The bottom extends about three-fourths of the distance back, leaving a space sufficient to allow the droppings to escape, and fall to the ground. There were three tiers of boxes one above the other—receding just sufficient to allow the open space to clear the top of the under box. The bottom should be bedded with clean straw, as fattening fowls should never be allowed to rest on the boards, and the straw should be often renewed.

It is recommended that the bottom, instead of a board, should be of narrow laths or rounded sticks, in order, more effectually, to allow the droppings to escape. They should be kept perfectly clean, and the fowls fed regularly three times a day. These coops should be under cover, and, if possible, at some distance from the other poultry, as the noise of their own species disturbs and prevents their thriving.

In Europe, many people, instead of putting fowls in pens, coop them up in cages hung in the open air, and made in such a manner that their heads peep out on one side, and their rumps on the other; thus packed up and immovable, they eat, sleep and digest nearly the same as in pens.

At the time when the French had a decided taste for spices and aromatics, people contrived the method of varying at pleasure the flavor and perfume of the flesh of poultry, by mixing up with the paste for fattening them sweetmeats of musk, aniseed, and other aromatic drugs.

It is well known that in partridges, when feeding on appletree buds, the flesh becomes bitter; it is also supposed that wild celery, on which the canvass-back duck feeds, in the Susquehannah and Potomac, gives them their exquisite flavor, so highly prized by epicures.

It has been said that some turkeys that had eaten a great many onion tops, had flesh of an exquisite taste; and that parsley, fennel, or garlic, introduced into the food for young turkeys, alters the flavor of their flesh for the better. What might not be expected by adding to the diet of fowls such substances as are capable of modifying the flavor of their flesh to advantage? The inference accordingly is fair, that if more careful experiments were made than have hitherto been done to ascertain the effects of different sorts of aliments or flavor, discoveries might be made of importance to poulterers, and all those who keep poultry.

The following remarks we find in an essay on fowls, published in the Journal of Agriculture:

"*Ever-laying Fowl.*—Any of the breeds of domestic fowls may become what is termed ever-laying, that is, not inclined to hatch, an artificial temperature produced by domestication.

"*Comparison of sorts.*—The things most necessary to attend to in selecting sorts for breeding, are the number, size and flavor of the eggs, and the color and delicacy of the flesh.

"*Number of eggs.*—The Scotch-Russian cross with the common fowl seems to stand highest, and next to these the Dorking, Poland and Spanish.

"*Size of eggs.*—The largest eggs are laid by the Poland and Spanish, the next by the Dorking; the game sorts and all the small varieties lay small eggs.

"*Flavor of eggs.*—The finest flavored eggs are those with the brightest yolks, such as are laid by the game breed, and by speckled varieties of the common breed. The large eggs of the Poland and Spanish have often pale yolks, and are deficient in flavor.

"*Color of the flesh.*—Those with dark colored or yellow legs have the flesh of a less pleasing white than those with pale flesh-colored, or white legs.

"*Delicacy of the flesh.*—The game breed, the Spanish, the Dorking, are the most delicate; the Malay coarse and inferior."

Selecting is a matter of some importance, since the quality of your poultry may be much improved by attention to this subject. Without attention to any particular variety we will enter upon some general remarks on this department. In choosing stock, select young fowls, and from such as have been remarkable for their good laying and thrift. After one season you will be able to select eggs from your own stock of such desirable qualities; they may be obtained by care and time, as well as any peculiarity of plumage you may fancy. The skill which the breeders of bantams and fancy pigeons obtain in giving such mixture of colors as they please to the plumage, is so great, that some have asserted they could breed to a single feather. Good fowls may be of any color, but to have none but good and handsome ones will require time, attention, and selection; and as they are no more trouble than inferior ones, there is no good reason why the farmer should remain satisfied with inferior fowls, than inferior cattle, sheep and swine. There is the same proportionable difference between some of our common non-laying breeds, and the Poland and Dorking fowls, as there is in the scrub and short horns, the shad-bucks, land-pikes, &c., and the Berkshires.

CHAPTER V.

POULTRY-HOUSES, ETC.

Necessity of a house and yard—situation—extent—advantages—should neither be wet nor exposed to winds—should be kept exclusively for gallinaceous fowls—plans—Lord Penryn's—Mowbray's—Queen Victoria's—the cottager's—Wakefield's—Scotch—cottages best—our own—Virginia—New Jersey—Rhode Island—poor man's and New York poultry-house.

WHETHER fowls are suffered to run at large, or are confined, there should be a poultry-house and yard where they may be regularly fed. Previous, therefore, to getting a stock of poultry, a place should be prepared for them. We will begin with the yard and house. In selecting a situation for this purpose it will be necessary to have them, if possible, on the south side of some building, or the south or south-east side of a hill or bank, so that one side of the wall may be set against the hill, and if of stone to be laid in mortar, which will add much to the warmth of the room. We would suggest, too, as an object of economy, when building the wall, to leave holes or recesses in them of twelve inches square, in which shallow boxes or drawers may be placed for the nests. These drawers can be removed when necessary, and cleansed or freed from vermin. If the buildings are of wood, they should be filled in with brick or lathed and plastered. Brick or stone is preferable, as they harbor less vermin.

The confinement of fowls will be found a most necessary arrangement, as on many occasions it is highly necessary they should be confined, as at planting time, or at some other periods when they are particularly troublesome: close confinement in a room or shed would soon interrupt their laying and make them sick, but a yard on the plan we are about to describe would answer every purpose, and be often found very advantageous in securing the eggs of such fowls as had contracted a habit of laying away from the house and barn, and endangering the loss of eggs.

It is well known that cold benumbs fowls, retards and diminishes their laying; and that a too intense heat diminishes their laying; that the want of good water gives them the pip, costiveness, and other inflammatory diseases; in fine, an infectious atmosphere makes them drooping, whence it naturally follows that their fecundity is less, that the flesh is not of so good a quality, and that the rearing of them is difficult. Under such circumstances one may judge how important it is for the improvement of poultry, that it should be wholesomely, comfortably, and cleanly housed.

Dickson says, "In order to unite all the advantages desirable in a poultry-yard, it is indispensable that it be neither too cold during winter, nor too hot during summer; and it must be rendered so attractive to the hens as to prevent their laying in any chance place away from it. The extent of the place should be proportional to the number of fowls kept, but it will be better too small than too large, particularly in winter, for the mutual imparting of electricity and animal heat. There is no fear of engendering infectious diseases by too much crowding; and it is found, in fact, that where fowls are kept apart they are much less prolific."

The driest and warmest soils are best adapted to the successful rearing and breeding domestic fowls, especially chickens; and to be attended with the greatest success and least trouble, some expense and great precaution will be required. Fowls endure severe cold much better than moisture. To unite all the advantages desirable in a poultry-yard, it should neither be wet nor exposed to cold winds. There should, if possible, be running water in the yard, and under cover should be placed ashes and dry sand, where they may indulge in their natural propensity of rolling and basking or bathing themselves. Gravel, broken shells, crushed bones, and old lime mortar, should always be placed within their reach.

From our own experience we are satisfied that the same house ought to be kept exclusively for barn-yard fowls; for though they will not be very disso ciable with others through the day, they do not like to sleep under the same roof with different species from themselves. Turkeys, in particular, are very quarrelsome, and will not suffer the other fowls to come near them. Geese, too, are troublesome at the feeding troughs by keeping the fowls away till they have satisfied their hunger; ducks soil the water, but are less troublesome than turkeys or geese.

Having settled all preliminaries, we will now proceed to give several plans, some of which would be rather expensive, and intended more particularly for the wealthy or fancy farmer; while others would be more simple and unpretending, and for utility rather than show, and could be erected at a very trifling expense, and within the reach of every one. From these plans any carpenter could erect one to suit the taste of the most fastidious.

We will therefore begin by giving some of the English and Scotch plans, and add several plans adopted in our own country, which differ in some points very materially from the foreign, and which will probably be found as well, if not better calculated for our purposes than either of them; and by having a description of the different kinds a choice can be made, or one constructed by taking parts of either and combining the advantages of the whole.

LORD PENRYN'S POULTRY-HOUSE.

This establishment is described by Dickson in his work on Poultry, as follows: " The most magnificent poultry-place perhaps that has ever been built, is at Lord Penryn's, at Winnington, in Cheshire. It consists of a handsome elegant front, extending about 140 feet; at each extremity of which is a neat pavilion, with a large arched window. These pavilions are united to the centre of the design by a colonnade of several cast-iron pillars painted white, which support a cornice, and a slate roof, covering a paved walk, and a variety of different conveniences for the poultry, for keeping eggs, corn, and the like. The doors into these are of lattice-work, also painted white, and the framing green. In the middle of the front are four handsome stone columns, and four pilasters, supporting likewise a cornice and a slate roof, under which and between the columns, is a beautiful Mosaic iron gate; on one side of this gate is an elegant little parlor, beautifully papered and furnished; and at the other end of the colonnade a very neat kitchen, so excessively clean, and in such high order, that it is delightful to view it. This front is the diameter or chord of a large semicircular court behind, round which there is also a colonnade, and a great variety of conveniences for the poultry. This court is neatly paved, and a circular pond and pump in the middle of it. The whole fronts towards a rich little field, or paddock, called the poultry-paddock, in which the poultry have liberty to walk about between meals. At one o'clock a bell rings, and the beautiful gate in the centre is opened. The poultry being then mostly walking in the paddock, and knowing by the sound of the bell that their repast is ready for them, fly and run from all quarters, and rush in at the gate, every one striving which can get the first share in the scramble. There are about 600 poultry of different kinds in the place; and although so large a number, the semicircular court is kept so very neat and clean, that not a 'speck of dung is to be seen.'

"This poultry-place is built of brick, excepting the pillars and cornices, and the lintels and jambs of the doors and windows; but the bricks are not seen, being all covered with a remarkably fine kind of slate from his Lordship's estate in Wales. These slates are closely jointed and fastened with screw-nails, on small spars fixed to the brick; they are afterwards painted, and fine white sand thrown on while the paint is wet, which gives the whole the appearance of the beautiful freestone."

"This sort of cleanliness," says Mr. Beatson, "with as free a circulation as possible, and a proper space for the poultry to run in, is essential to the rearing of this sort of stock with the greatest advantage and success; as in narrow and confined situations, they are never found to answer well."

MOWBRAY'S POULTRY-HOUSE.

"Whether or not the poultry be suffered to range at large," says Mowbray, "and particularly to take the benefit of the farm-yard, a separate and well fenced yard or court must be pitched upon. Upon farms the poultry-yard may be small, as the poultry should be allowed to range over the premises, to pick up what cannot be got at by the swine. The surface must be so sloped and drained as to avoid all stagnant moisture, most destructive to chickens. The fences must be lofty and well secured at the bottom, that the smallest chicken cannot find a passage through, and the whole yard perfectly sheltered, from the north-west to the south-east. It should be supplied with some effete lime, and sifted ashes, or very dry sand, in which the fowls may exercise the propensity, so delightful and salutary to them, of rolling or basking themselves. This is effectual in cleansing their feathers and skin from vermin and impurities, promotes the cuticular excretion, and is materially instrumental in preserving their health."

The poultry-house within the yard, if there be a choice, should have a southern aspect, defended from cold winds and the blowing in of rain or sleet.

If the number of stock be considerable, the houses had far better be small and detached, both for health and safety sake; and especially they should be absolutely impenetrable to vermin of any description; should these houses abut upon a stable, brew-house, or any conductor of warmth, it will be so much more comfortable and salutary to the poultry.

Fig. 1.

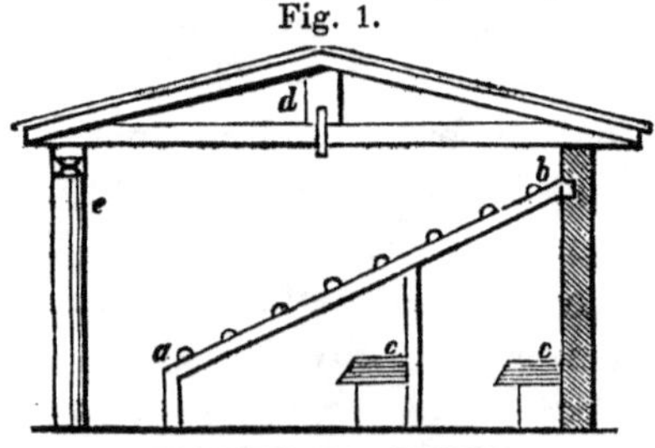

TRANSVERSE SECTION.

The elevation should be a simple style, and, for health's sake, the roof should be lofty; the perches will be more out of the reach of vermin, should any break in; and there should be only one long and level range of perches, because, when they are placed one above another, the fowls dung upon each other; convenient steps driven in the walls, will render easy the ascent of the poultry to their perches; or they may be made as shown in the above figure, in which *a b* are spars for the poultry to sit on; *c c*, ranges of boxes for nests; *d* the roof, and *e* the door, which should be nearly as high as the ceiling for ventilation, and should have a small opening with a shutter at bottom, to permit the poultry to go in and out at pleasure. The spars on which the birds are to roost should not be round and smooth, but roundish and roughish like the branches of a tree.

QUEEN VICTORIA'S POULTRY-HOUSE.

(*See Frontispiece.*)

"In a secluded wood on the boundaries of the Home Park, stands the HOME FARM or the farm attached to Windsor Castle—the private farm of her Majesty. In this establishment, which was founded by George III., are situated the royal fowl-house and poultry-yards, but of which, notwithstanding their great interest, the public know nothing, save the mere fact of their existence. Here, her Majesty, retiring from the fatigues of state, finds a grateful relief in the simple pursuits of a country life.

In cultivating the homely recreations of a farm, ner Majesty has exhibited great industry and much good taste. The buildings and farm routine which sufficed for the clumsy management of 1793, have been discovered by her Majesty to be totally unsuited to the more enlightened system of 1843, and hence, under the direction of her Majesty and Prince Albert, and others, an entire reorganization of the establishment has been determined, and is now in progress.

"The fowl-house, lately built at Windsor, is a semi-gothic building, of simple and appropriate beauty. It consists of a central pavilion, used for inspecting the fowls, crowned on the top by an elegant dove-cot, and on the sides, of wings capable of symmetric extension, in which are placed the model roosting houses, and laying and breeding nests of the fowls. The ground, in front, slopes towards the Park, and is enclosed and divided by light wire fences, into separate wards, for the "run," or daily exercise of the birds. Inside these wards, gravel walks, bordered by grass plots, lead to the entrances of the fowl-house. In the proportions, distributions, and fittings of the apartments of this house, considerable knowledge of the habits, with a corresponding and most commendable regard to the conveniences of their granivorous tenants, has been displayed; the chambers are spacious, airy, and of an equal and rather warm temperature, which accords with their original habits, and their nests are made as far as possible to resemble the dark bramble-covered recesses of their original jungles. In this particular, her Majesty has set a good example to the farmers, who too often follow the false routine of their fathers, rather than consult the habits and obey the natural instincts of the animals about them."—*London Pictorial Times.*

THE COTTAGER'S POULTRY-HOUSE.

Boswell says, "No housewife in the country, whether her cottage is situated in a village or on the roadside, need be without a few fowls; provided proper care be taken to do no injury to surrounding property not her own, and of whose possessor she has not obtained permission. This can be more easily accomplished even in apparently disadvantageous circumstances, than is generally imagined. Although in some cases the profit may appear small, in all, the economy is great. Suppose a cottager who has taken up house in the country, how desirable is it that he, while engaged in his daily toil, and his wife in her usual avocations, should have some means by which even the scraps of their scanty table, without entrenching upon their time, should be turned to profitable account, instead of littering the floor, or being swept out into the road!

For the accomplishment of this most desirable object, the means, in the great majority of instances, are simple and easy.

At the gable end of the house, as near as possible to the opposite side of the kitchen fire—at which part, and for this purpose, the wall might be made thinner, perhaps only brick in bed, or brick on end, to the height of six or eight feet—a poultry-house, something similar to the annexed figure, might with the slightest materials be made. Formed of rough slabs, or of such other material as the rural resident must be at all times able to find in his neighborhood, such an erection would cost almost no expense. Its construction can easily be understood from the figure. Its size, form, and fitting up must depend principally upon the judgment and convenience of the rearer; but in such a place, it not being advisable to permit them to roam at large, a palisade can easily be formed completely to enclose this poultry-house. It will only require one row of stakes, running parallel with the gable of the house, with the shed at one end and a small gate at the other, to form a complete parallellogram, in which the poultry, when necessary, may be confined. The enclosure does not require to be high, if the tops are pointed, for these fowls seldom attempt to pass over. If this enclosure can be made to surround the ash-pit, even in very unfavorable circumstances, a very perfect and profitable poultry establishment may on a small scale be formed.

Fig. 2.

In this hen-shed, proper sticks for roosting will of course be fixed, with a space to be permanently left open during the day, although in it there should also be a small door, to be constantly shut at night; the whole upper division should be upon hinges, for the purpose of cleaning and inspecting more easily. In the lower section and division there should always be kept a proper supply of soft short straw, as it will be found that the hens will frequently give a preference to such a situation for laying; but the back part of the upper division should be divided into similar apartments, to give them as much variety as possible in the choice of a nest. The lower part, if well furnished with straw, will form a very tolerable receptacle for a few ducks, while it will always be an excellent refuge for the other fowl, in a rainy day. If one-half of the lower division next the house, and at the place where the fire-place is, be boarded up, it will form for them a very snug berth, even in the coldest weather.

It need scarcely be remarked, that the larger the enclosure the better; for, although it has been so much practised in France, and so warmly recommended by French writers, we are no advocate for too strict confinement, unless it be from rain and damp. When the state of the fields cannot render it injurious, they should at least occasionally be allowed to roam at large. At the roots of hedges they will grub up many a hearty meal, and the public roads will furnish them with a more grateful rolling bed than all the artificial mounds and hollows of brick-dust, sand, and ashes, which can be laid down in the poultry-yard.

In rearing our simple shed, we have taken it for granted that the situation is completely dry. This, we apprehend, will be the case in a hundred instances or two; but when it is not, proper care should be taken to render it so, either by draining, sloping, or elevating."—*Boswell.*

MR. WAKEFIELD'S POULTRY-HOUSE.

It is well said by Mr. Beatson, that poultry, when rightly managed, might be a source of great profit to the farmer; but, when many are kept, they ought not to be allowed to go at large, in which case little profit can be expected; for not only many of their eggs will be lost, and many of themselves perhaps destroyed by vermin, but, at certain seasons, they do much mischief both in the barn-yard and in the field. No doubt they pick up some grain at the barn-doors, that otherwise would be lost; but if the straw is properly threshed and shaken, there would be very little of this waste. In the common careless way of threshing, a great deal of corn is undoubtedly thrown out among the straw; but, when we consider the dung of the fowls and their feathers that get amongst it, and the injury these may do to the cattle, this is no object, and the poultry should never be allowed to scrape or roll among the straw at all. Some people are of opinion, that each sort of poultry should be kept by itself. This, however, is not absolutely necessary; for all sorts may be kept promiscuously together, provided they have a place sufficiently large to accommodate them conveniently, and proper divisions and nests for each kind to retire to separately, which they will naturally do of themselves.

This mode has been practised with great success by Mr. Wakefield, near Liverpool, who kept a large stock of turkeys, geese, hens, and ducks, all in the same place; and, although young turkeys are in general considered so difficult to bring up, he reared great numbers of them every season, with little trouble or difficulty. He had about three quarters or a whole acre enclosed with a fence only six or seven feet high, formed of slabs set on end, or any thinnings of fir or other trees, split and put close together, fastened by a rail near the top, and another near the bottom, and pointed sharp, to prevent the poultry flying over, for they never attempt this, although it is so low. Within this fence were places done up slightly, but well secured from wet, for each sort of poultry, and also a pond or stream of water running through it. Mr. Wakefield's poultry were fed almost entirely on potatoes boiled by steam, and thrived astonishingly well. The quantity of dung that is made in this poultry-place is also an object worth attention; and when it is cleaned out, a thin paring of the surface is at the same time taken off, which makes a valuable compost for the purpose of manure.

For keeping poultry on a small scale Mr. Beatson thinks it is only necessary to have a small shed or slight building, formed in some warm, sheltered situation (if near the kitchen or other place, where a constant fire is kept, so much the better), fitted up with proper divisions, boxes, lockers, or other contrivances, for the different sorts of birds, and for their laying in.—*Communications to the Board of Agriculture*, (*Eng.*)

SCOTCH POULTRY-HOUSE.

In a paper published in the "Transactions of the Highland Society of Scotland," for 1833, Mr. England, of Aberdeen, gives a plan of a poultry-house, which appears to be very complete of its kind, though he differs from most of the authorities in many points of management. The house is divided into separate wards, each ward fitted up to lodge twenty-four hens and one cock, with a yard attached, about fifteen feet long, by ten broad. The annexed is a ground plan of two such wards, with their yards and houses.

Fig. 3.

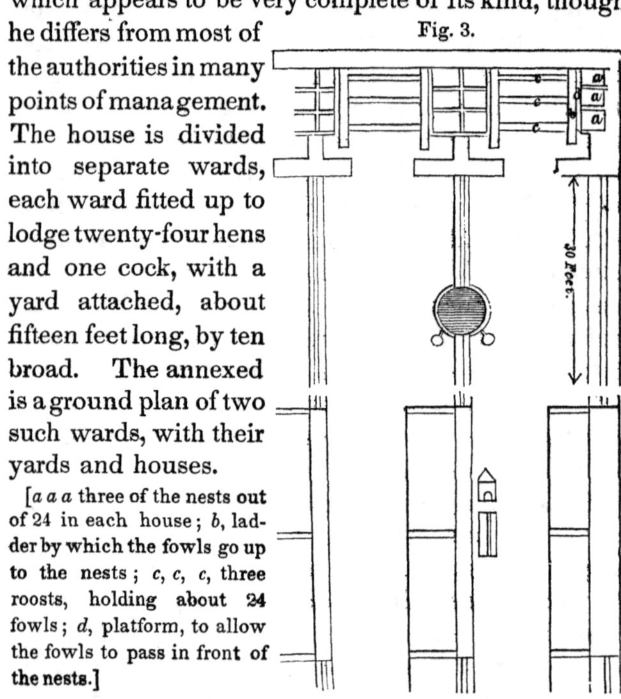

[*a a a* three of the nests out of 24 in each house; *b*, ladder by which the fowls go up to the nests; *c*, *c*, *c*, three roosts, holding about 24 fowls; *d*, platform, to allow the fowls to pass in front of the nests.]

The manner in which the nests, the roosting perches, the ladders for the fowls to go up by, and the platforms to allow them to pass in front of the nests, are arranged, will be best understood from the following Figures.

Fig. 4.

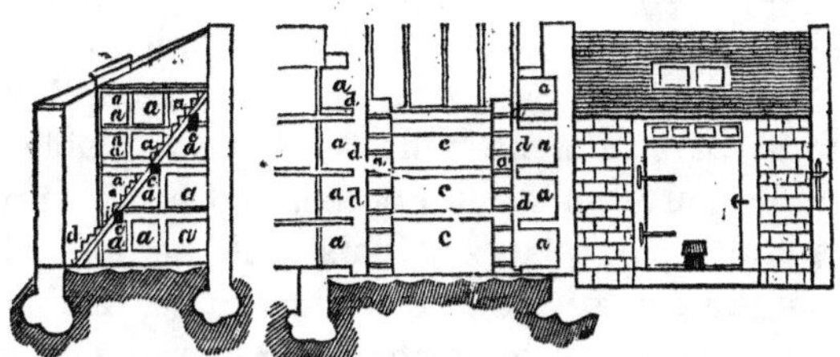

Mr. England also provides what he terms a Storm house, for the fowls to run for shelter in bad weather, and a dry bath-house to afford them sand for pulverising, which are represented in Figure 5.

Fig. 5.

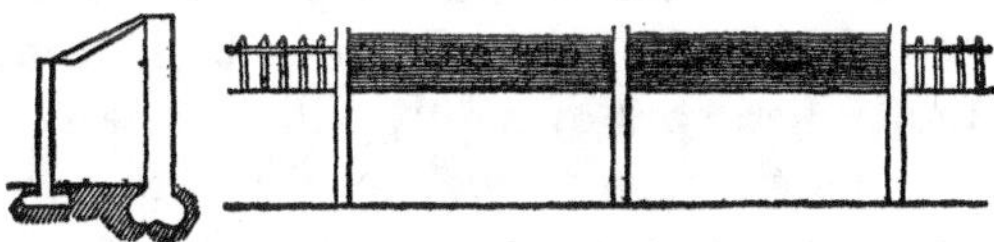

STORM-HOUSE.

THE COTTAGER'S BEST.

Always in the building of a cottage, and sometimes even when there was no intention of the kind when it was built, very ample accommodations for poultry can be provided, without almost a shilling of additional expense. To this purpose a part of the space next the roof (garret), so often unoccupied and useless, might easily be devoted. To accomplish the object, a part of it next the kitchen-fire gable end should be partitioned off, floored, and fitted up with roosts and laying places. This could be done either on a large or small scale, according to the inclination or the means of the projector. An opening of sufficient width should be made in the wall, at the height of the lower ceiling, through which the fowls could be conducted, by means of a *hen-ladder*, to the enclosure prepared for them below. Of course there must be a hatchway somewhere, to afford access for the purpose of inspection and cleansing; for the proper position of which the builder or possessor can be at no loss. If the attics are sufficiently high, it may be placed anywhere, but evidently with greatest convenience in the passage of the house; but if they are low, the nearer the space partitioned off for the reception of the fowls the better.

This is a location for poultry possessing many advantages. Having their berth immediately above the kitchen, they are secured in a proper degree of dryness and warmth, which, in winter, especially with the spring-hatched pullets, will tell well in the production of eggs. Perhaps this is the best hen-house locality for securing eggs in winter which can be suggested to the frugal and judicious. Besides, the fowls are here free from many dangers, and safe from many enemies, to which they are exposed in a lower and more open situation. There may be rats and mice in a cottager's dwelling, from which even in such a situation as this his poultry may not be secure, and by them their eggs may be destroyed; but there are many reasons, besides the preservation of his poultry, why these should, to the utmost possible extent, be banished. We must therefore consider this evil as non-existent, and from every other his feathered brood here are safe. The weasel, the pole-cat, and fox, may prowl around in vain; and what is of still more importance, he will avoid a danger to which we have more than to all the others put together been exposed, and by which we have been annoyed and suffered more—he will be safe from the midnight pilferers, who know so well how to rob the hen-roost mercilessly, in secrecy and silence, and who do not know the value of a good fowl even after it is stolen, but who would treat the best bantam, Jungle, or Dorking, as remorselessly as Meg Merrilies, who had more respect for the savoriness of the mess, than the nature of the ingredients, or the manner of procuring them.—*Boswell.*

In the foregoing, we have given several plans and descriptions of foreign poultry-houses; we will now proceed to give some American, among which will be found some of our own plans. If there is nothing original in them, they are none the less useful.

Fig. 6.

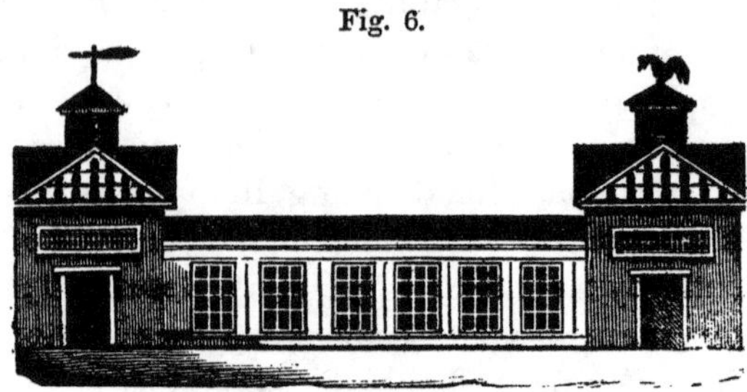

OUR OWN POULTRY-HOUSE.

The above figure represents the front and elevation of rather an extensive and costly establishment, which would be very convenient, and add much to embellish the premises. The buildings at the ends are intended for laying, hatching, and roosting apartments. The cupolas on the tops are finished with blinds, for the purpose of ventilation as well as ornament. On the bottom of the cupolas, and inside of the building, should be a door, hung on hinges, with a cord attached passing through a pulley, so that it may be closed or opened at pleasure, to ventilate when required. In the gable ends, if facing the south, dove-cotes may be formed; or they may be made in the roof, as in the figure.

The long building with windows in front, connecting the two extreme ones, is intended for a *storm-house*, *chicken saloon*, or *walk*, for exercise in the winter, as well as a retreat from storms, feeding, basking, gravel, sand, lime, &c.; being made warm by filling in with brick, or lathed and plastered, and the roof should be thatched with straw. The front should be ten feet high, roof sloping to the north. The windows are intended to admit heat in winter, as well as light. If only for a storm-house, the windows may be omitted, and the front finished in the form of a shed. It will be found very convenient for feeding and watering, as well as for gravel, ashes, lime and sand. Boxes for nests may also be placed there for laying and hatching. By partitioning it off, two varieties of fowls may be kept separate; or one side may be appropriated for turkeys and guinea-hens, and the other to the gallinaceous tribe. Doors from each should open into the yards, which should be of considerable size; say, at least half an acre for two hundred fowls, as room and space in the air is necessary for their health, when they are not permitted the range of the barn-yard. The yard should, if possible, be a little sloping, that it may be dry, as moisture is a most destructive enemy to poultry. It should be enclosed by a fence, at least seven feet high, with long, sharp pickets, and the timbers on which the pickets are nailed, unless some distance below the top, should be on the outside, to prevent the fowls perching on them, as they seldom attempt to fly over a fence without alighting. When first confined, if they have been used to roam over the premises, they will show some impatience, which soon wears away, if everything else is made agreeable to them. It may, however, be necessary to clip the wings of some of them, when first introduced, particularly if taken from the barn-yard, where they have always had their liberty. My fowls are so attached to the yard that they are unwilling to leave it, even when the gates are open—the effect of education.

The buildings at the ends should be thirteen feet square, and thirteen feet posts. We name this size, as there would be no waste of timber, being just the length of the boards and joists. If not too near the dwelling-house, so that there would be danger of fire from sparks, we would recommend to have the roof thatched with straw, as being much warmer in winter, and cooler in the summer, and when well done, it forms a light and durable roof, and will last for twenty years. It should, however, be made very sloping, in order to carry off the water more readily. A hole one foot in diameter, about two feet from the ground, with a door, either to slide, or hung on hinges at the top, which may be held open by means of a cord, should be made in each department, for the fowls to pass in and out, and to confine them when necessary. There should be no floor in the first story to prevent the fowls from coming to the earth; and the litter should be often removed, and the bottom sprinkled with lime, at least once in each week.

In the second story, there should be a tight floor under the roosts to catch the droppings of the fowls, by which means the apartment can be kept much cleaner, and the manure may be saved and sold to the morocco-dressers, for which they readily pay eighteen cents per bushel; or it may be gathered for manure, which, with the exception of pigeon's dung, is said to be the strongest of all animal excrements—*it is home-made Guano.* This will add an item to the profit of keeping fowls, that has heretofore been entirely overlooked.

The roosts should commence on one side, at the top, near the plate, and slope downwards, at an angle of about 45 degrees, like a ladder, to within eighteen inches of the floor. The spars for the roosts should be about two inches square, with the corners taken off, and placed eighteen inches apart horizontally, for the fowls, and at least two feet for turkeys, so that they may not incommode one another by their droppings. No flying is necessary in this form of a roost, as the birds ascend and descend by steps. It is recommended by some writers, to have the spars or cross-pieces for roosting, of sassafras, two inches in diameter, with the bark on, which is said to be a protection from vermin—in which, however, we have no faith.

The lower story is designed for the laying and hatching apartment. When we first erected our poultry-house, we tried ranges of boxes, similar to those generally made for pigeons, placed against the sides of the walls for nests—but experience, the best of teachers, proved it was erroneous, especially when hatching; for when the setting hen left her nest to procure her food, drink, &c., one of the other hens would espy the eggs, and pop in and lay her egg. In the mean time the hatching hen would return and find her nest occupied; and finding it no easy matter to eject the intruder—as possession, with hens, like men, is considered nine points of law—would seek the first nest she could find with eggs, and settle herself there very contentedly. The consequence was, the other hen, after depositing her egg, would leave the nest, and the eggs would cool and spoil. There is another difficulty. If vermin should make their appearance, there is no way of getting at them or cleansing the nests; to remedy this, we had separate boxes made and hung around the sides, and placed in the corners, which can be removed when hatching, or cleaned and freed from vermin when necessary.

The best arrangement for nests which has fallen under our notice, was at the seat of a gentleman residing near New Brighton, on Staten Island. In one end of the building, and extending about one quarter of the length, a platform is raised about five feet high, on which a frame or rack is erected, and divided off into sections of about eighteen inches square. The nests are made in the form of a shallow drawer, and are inserted into every other square in the frame, and facing the wall. By this arrangement the hen is not annoyed by the other fowls, nor disturbed by persons entering the house.

If they become foul, and require cleansing, they are easily removed.

Since the foregoing was written, we have seen some account of the poultry establishment of Queen Victoria, at Windsor, in which the laying nests are composed of dry twigs of heather and small bramblets of hawthorn, covered over with the lichen. These materials, rubbed together by the motion and pressure of the hen, emit a light powder, the produce of the crushed leaves; and this, finding its way between the feathers to the skin, was found to have the immediate effect of discharging the bird of every description of parasite. This was accompanied with a drawing representing the nests, of which the figure below is a copy.

Fig. 7.

From the above we took a hint, and procured some hemlock boughs, and tacked to our box nests, nearly obscuring the entrances, giving the hen an opportunity of gratifying her propensity of secretiveness. This arrangement seemed to be very satisfactory to the hens, and added much to the appearance of the room.

The size and shape of the yard may be made to suit the convenience and taste of the owner, but from our experience, the larger the better, particularly if there are more than one sort of fowls kept together. Turkeys and geese require a greater range than the gallinaceous fowls. Sheds should be erected for shelter from storms in summer, and protection from cold in winter. A hedge of lilac, or any other sort of shrubs within the fence, or what are better, small evergreen trees, with the limbs left as near the ground as possible, will be found very acceptable to the poultry, where they will retire for shelter from the heat of the sun, and it is a good protection and screen from the view of the hovering hawk. Next to evergreens we would recommend the mulberry and cherry, as they are very fond of the fruit.

VIRGINIA POULTRY-HOUSE.

The following plan for a poultry-house seems well calculated for a warm climate. It was communicated to the Editor of the Cultivator, and published in Vol. V., page 47. The writer says, "I have used the poultry-house of which the following drawing is a representation, for about eight years, and can testify that it is preferable to any known in this section of country, and many of my neighbors have thrown away their old houses and built after my plan.

"The roosts for the fowls should be often renewed, and always of sassafras, as the smell of that wood is deleterious to the vermin on poultry. The floor in the sitting-room should be kept sprinkled with lime, and the litter from under the house taken away weekly."

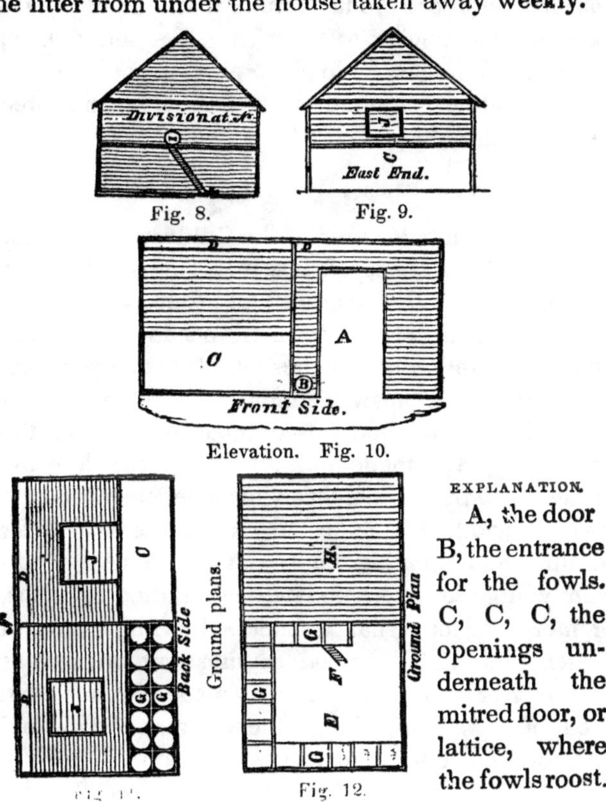

Fig. 8. Fig. 9.

Elevation. Fig. 10.

Ground plans.

Fig. 11. Fig. 12.

EXPLANATION.

A, the door
B, the entrance for the fowls.
C, C, C, the openings underneath the mitred floor, or lattice, where the fowls roost.

D, D, D, six inch openings to admit air. E, the ground floor, made of earth, elevated above the surface one foot, with boxes for the poultry to lay and sit in, to be increased with tiers as they may be required. F, steps for poultry to go to roosting room. G, G, G, G, boxes for nests. H, lattice, or mitred floor for the litter from the poultry to fall through, and room for the poultry to roost in. I, a round hole, one foot in diameter, for fowls to pass to roost. J, J, J, lattice windows or blinds three feet wide, and three feet six inches deep.

NEW JERSEY POULTRY-HOUSE.

In the seventh Volume of the Cultivator, Mr. Van Uxen, of New Jersey, says, "Having made some experiments in the raising of chickens, a business that forms a part of every farmer's occupation, I send you a description of my present plan of operation, which appears to answer admirably. Under an outhouse 16 by 18 feet, raised three feet above the ground, I have made a cellar three feet below the ground, making the height six feet altogether. Eight feet in width of this cellar is partitioned off for turnips, the remaining ten by sixteen feet being sufficiently large to accommodate one hundred chickens, or more. This cellar is enclosed with boards at present, but it is intended to substitute brick walls in a year or two. The roost is made sloping from the roof to within eighteen inches from the ground or floor, twelve feet long by six feet wide. The roost is formed in this way: Two pieces of two-inch plank, six inches wide, and twelve feet long, are fastened parallel, six feet apart, by a spike or pin, to the joist above, the lower end resting on a post eighteen inches above ground. Notches are made along the upper edge of the plank, one foot apart, to receive sticks or poles from the woods, the bark being left on. When it is desirable to clean out the roosts, the poles, being loose, are removed; the supports, working on a pivot, are raised and fastened up, when all is clear for the cleaning out. I next provide the fowls with corn, oats, and buckwheat, in three separate apartments, holding about half a bushel each, which are kept always supplied. They eat less, I find, if allowed to help themselves to what they want, than if fed to them in the usual way, for, in the latter case, each tries to get as much as it can, and thus burdens itself; but finding in the former case that they have abundance, they eat little, and that generally in the morning early, and in the evening just before going to roost.

"I have sixty fowls, and they eat about six quarts per day of the three kinds of grain, in the proportion of twice as much corn as buckwheat or oats. In the roost is placed a trough of water, renewed every other day, burned oyster-shells, shell marl, and ashes.

"A row of nests is constructed after a plan of my own, and does well. It is a box, ten feet long and eighteen inches wide; the bottom level, the top sloping at an angle of forty-five degrees, to prevent the fowls roosting on it; the top opens on hinges. The nests, eight in number, are one foot square; the remaining six inches of the width is a passage way next to the wall, open at each end of the box, and another opening midway of the box; the advantage is to give the hens the *apparent secrecy* they are so fond of.

"The following are the advantages of this plan of keeping fowls. By having a roosting-place partly under ground, the fowls can keep warmer through the winter than any roost above ground could be made without fire; and this is absolutely necessary to induce them to lay. I would recommend this plan to all who wish to make their fowls profitable."

RHODE-ISLAND POULTRY-HOUSE.

The following plan of a poultry-house is also taken from the Cultivator, Vol. VIII., which differs some from those already given. The writer who furnishes the plan remarks, "Some farmers are of an opinion that a few boards tacked together, or set against the side of a wall, answer very well for the purpose of a hen-roost; but I have come to the conclusion that to render our fowls profitable, as much care must be taken of them as of our horses and cattle. This house may be built of pine boards, or it may be clapboarded and plastered with lime; in either case it should have a good plank floor. It is twelve feet long, eight feet wide, and seven feet high from the bottom of the sill to the top of the plate.

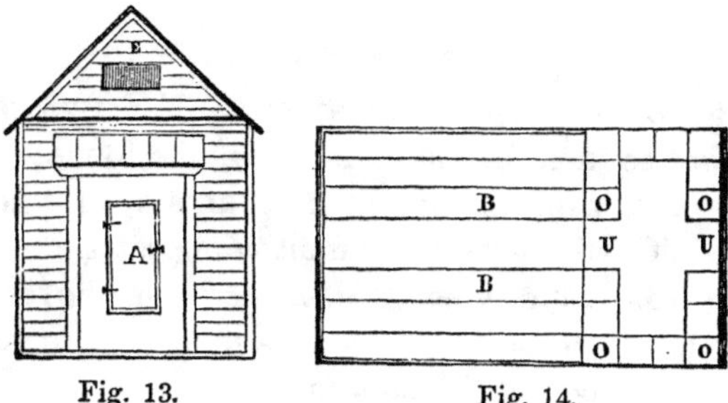

Fig. 13. Fig. 14.

EXPLANATION.

Fig. 13. View of the east end; A, a door, two feet wide and five feet high; E, a small window for ventilation.

Fig. 14. Interior view; U, a door; OOOO, boxes for nests, twelve inches square, to be placed in three tiers, one above the other; U, an inside door of the same dimensions as the outer one; BB, are poles, or roosts; these may be either sassafras or wild cherry-tree. They are fitted to swing up and hook at the upper floor.

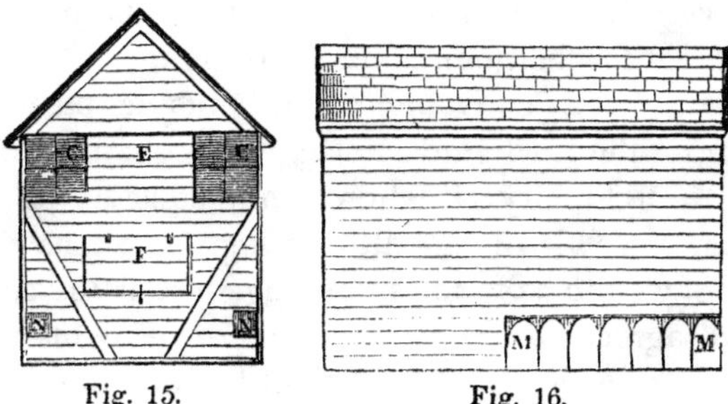

Fig. 15. Fig. 16.

EXPLANATION.

Fig. 15. View of the west end; NN, two holes, one foot square, for the entrance of the fowls; F, a door to throw out the manure; it turns up and hooks at E; CC, windows with small wire grates.

Fig. 16. Side view; MM, nests or boxes for brood hens; these should have a long door to swing down and hook at the bottom.

POOR MAN'S POULTRY-HOUSE.

A very cheap and economical plan of a poultry-yard and pen, is given by D. F. Ames, in the "Farmer's Rural Library." "When necessary to fatten any fowls for the table or market," says Mr. Ames, "the yard plan is far better than confinement in a dirty coop, where they generally first lose flesh, and afterwards contract a flavor by no means pleasant. One of these pens of the most simple form, and such a one as any handy lad could make in a few hours, should be attached to every cottage; it costs nothing but a very little labor, and would really be pleasant enjoyment for the noon hours, or evening.

"First, let a convenient and suitable place be chosen for the fowl-yard; not in a dark shady corner, but in a light, airy situation; and, considering the number of fowls intended to be kept, mark its size: it is not well to have too many together, as the cocks will disagree. A stock of twenty-five, containing two or three cocks, is sufficient for one house; if more are to be kept, erect another pen in a different direction; accordingly, mark out a place in the form of a circle of eighteen or twenty-four feet in diameter. On the outside of this circle, cut a trench three or four inches wide and deep, and plant poles twelve or eighteen inches into the ground every two feet. These poles should be as thick as a man's arm, and eight or ten feet high, thus forming a circle of poles

Fig. 17.

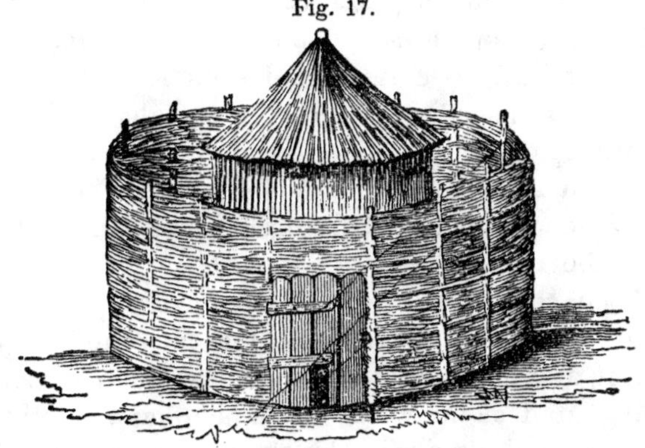

standing on end. Choose a space to the south, between two of the poles, for the purpose of a door, and the poles on each side of this space should be straight, and a little stouter than the rest; then go to the swamp or brush-wood, and cut a good parcel of it, leaves, small twigs, and burrs, all just as it stands. It ought to be six feet long, that it may reach three of the poles, and if longer all the better; then having conveyed it to the standing poles, commence by lacing some of the stout and straight ones round the poles in the trench, alternately in and out, like basket-work, going the whole round, the door-way of course excepted. When you have got it eight or ten inches high, stamp it well down, making all tight and firm, that the smallest chicken may not be able to pass through it. Go on thus till you get it five feet high, then pass the circle of brush over door-way and all, to make it firmer and stronger, continuing it up to the height of eight or ten feet; the upper may be lighter, and not braided so close; braid sometimes on one, and then on the other side of the uprights. Upon this principle, a yard may be made of any size, and in any situation, for really nothing. Any boy can make a door for this, and fix it with hinges from the sole leather of an old shoe.

"Then comes the fowl-house; this should be placed in the centre of the circle, that no vermin may get at it, and that the fowls may find shade and shelter all around, as the wind or sun may happen to be. A few stakes, a little more brush, and an arm-full of straw for thatch or roof, will make this answer; but one formed of boards with a good tight straw thatch, would be far preferable. Mind, I say, '*straw thatch*' for roof, as it is far the best thing; and if properly done, it will last twenty years. The sun, rain, and snow, have no effect on it. It is very warm in winter, and lets no heat through in summer. It should be formed of good clean, long straw, clean-threshed, and as little broken as possible; wheat or rye is preferable : put it on ten or twelve inches thick; I have seen it eighteen inches. Tie it close and securely with strips of white oak or hickory bark well twisted; but this every one knows how to perform. Mind and let the roof have a good pitch, or in other words, be very steep, that snow and rain may be quickly thrown off. To make this warmer in winter, the sides, either outside or within, may be laid with cedar brush and salt hay tacked up to the boards; or made of brush wicker-work, and then plastered outside and in with clay and short salt hay; and when dry, a good coat of lime white-wash. This gives a neat pretty look, and is warm and cheap. To have eggs all winter, hens must have a warm house; some, therefore, may prefer the following plan:

"Choose a location on the side of a hill, facing the south if possible, and forming an oblong of the size required; dig it down five or six feet, throwing the earth on each side; then some strong chestnut logs must be laid across, and a little brush on them; cover them with sods, and then with earth, and then good green sod on the top of all. The sides may have stakes driven in and boarded, or if the soil is sufficiently stiff, they may be omitted; do not neglect to fix a door. I have found a small hole, sufficiently large to let the fowls pass through, very convenient, when cut in the door and guarded by a board; you can easily count them when letting them out, or keep back any you may wish to catch. It is absolutely necessary that the house should be dry, and free from the effects of heavy rain on roof or floor. A small hovel may be easily constructed under ground, in a flat surface, sufficient for the accommodation of a cock and five or six hens; but care must be taken that the artificial mound do not furnish a sufficient elevation to assist the fowls over the fence; the precaution of cutting one wing will of course be taken.

"Now for the inside of the house—this should be arranged with order and neatness, for neglect on these points will be a serious evil; first, the nests—these may be formed by placing boards sixteen inches apart, beginning from the ground (do not floor it); then divide them into boxes by partitioning every sixteen or eighteen inches, and partly blind the front by nailing a board against them, leaving just room for the hens to pass out and in; a little piece of shingle can be placed across the bottom of the entrances to prevent the eggs from rolling out, and a perch can be so placed along the front as to assist them in getting up and into them. Choose those on the ground to hatch in, as the earth retains the temperature of the eggs better than hay or straw. Little doors would be convenient to place before the sitting hens to prevent their being disturbed.

"Perches for their nightly accommodation and roosting should be placed across so as not to have them dirty on one another, and down into the nests; and they should also be placed at different elevations, so as they can easily get up and jump from one to another. The perches or roosts should be of a good size, round, and stout as a man's wrist or arm, to make them steady, and to prevent their contracting the deformity of a bent or a crooked breast-bone, which is very common from this cause; they should also be so far apart that the fowls cannot from one perch peck those of another. Some fowls have a trick of doing this, and I have had several instances of the hens being almost stripped of feathers on the head and neck from others they did not agree with, and yet they would pertinaciously adhere to the situation that subjected them to the painful operation."

The hen-house should never be much larger than sufficient to accommodate the number of fowls to be kept in it; for if too large they huddle together in one corner, and, as it has been before observed, hens produce eggs more abundantly in a small apartment, than in a more spacious building. But warmth and cleanliness should be particularly attended to, and it should be rendered in every respect comfortable and agreeable to the birds that inhabit it; for, if that be not done, they will seek to lay away from home instead of in the nest provided for them, and if they cannot succeed, they will to a certainty produce fewer eggs, than if their propensities and tastes were better indulged; but if they have a clean, quiet, warm place to retire to, they will lay regularly and abundantly, and will repay both the trouble and expense.

Fig. 18.

NEW YORK POULTRY-HOUSE.

After detailing the conveniences and manner of construction of several establishments, we come now to a very simple, complete, and to our mind, a very efficient fowl-house, as given by a correspondent under the signature of H. in the American Agriculturist. The writer says, " The accompanying plan and references render a particular description unnecessary. The north, east, and west sides of the house are of brick; the floors are of cement to keep out rats.

" Fowls will not lay well in winter, unless they have during the day a dry, light, and warm apartment in cold and stormy weather. The room marked C is designed for this purpose; it is lighted in front and above by sashes, one of which, in front, is hung with hinges for the entrance. If necessary, a ventilator may be added to the roof, or a window in each end."

Fig. 19.

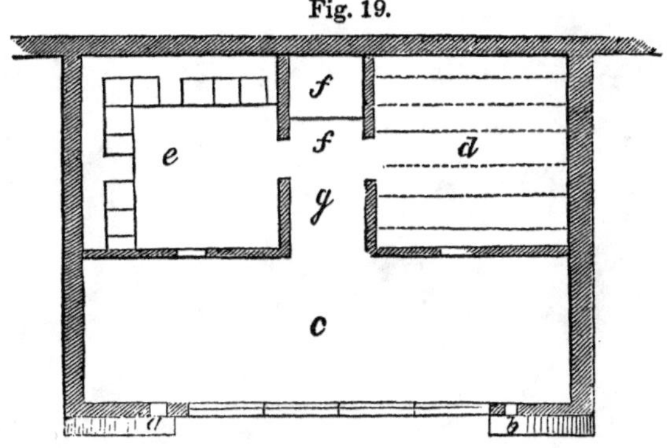

Fig. 20.

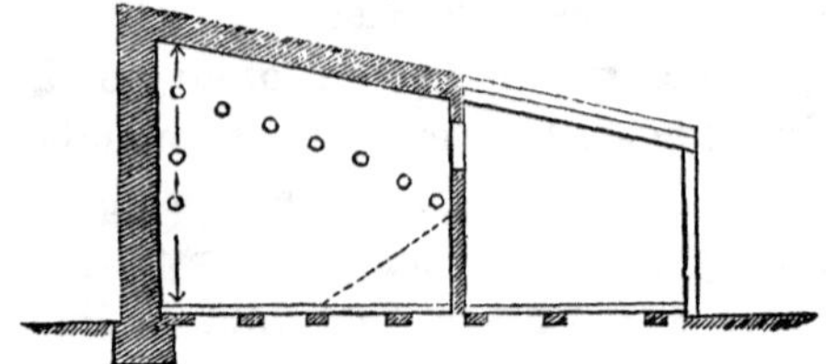

EXPLANATION.

Fig. 18. Elevation.

Fig. 19. Ground plan—*a*, *b*, apertures for admitting fowls, with slides for closing; *c*, place for feeding; *d*, roosting-room; *e*, laying-room with secluded nests; *f*, bin for feed; *g*, passage.

Fig. 20. Section through *a*, showing the position of the roosts.

CHAPTER VI.

FURNITURE AND ACCESSORIES TO THE POULTRY-HOUSE.

Furniture and accessories—Scotch feeding-hopper—Bement's hoppers—water-fountains—plans for nests—coops.

1st. In the place appropriated for the newly hatched chickens, there ought to be separate cages or coops, where each mother remains eight days with her family, after which she is removed into an enclosure to finish the rearing of them, till they can without danger be trusted by themselves.

2d. A small trench or low box filled with dry sand and ashes, in which the fowls may roll to clean their feathers and free themselves from vermin.

3d. Another trench containing horse-dung, to be frequently renewed, and in which they amuse themselves, particularly in winter, by scratching for grain and worms.

4th. A plat of grass on which they may pasture and divert themselves.

5th. A thick bushy hedge, or what are better, trees to furnish shelter from the heat of the sun, and screen them from the view of the hovering hawk. The best sort of trees probably are the mulberry and the cherry, as fowls are fond of the fruit; and we have found evergreens, such as the white pine, cedar, balsam, and termerack, very acceptable to them in a windy or hot day.

6th. A shed, even if low and narrow, where they can take shelter from rain.

7th. If there is not running water in the yards, stone or wooden cisterns, troughs, or water-fountains (see Fig. page 123) with pure water, in order to prevent their seeking by chance what is bad or corrupted.

8th. The yard, in short, should be dry, spacious, and kept as neat and clean as possible.

9th. There should be a number of coops, which can be moved (see Fig. page 126), to secure the hens and protect them from cold and rain, as well as from skunks.

10th. Feeding hoppers (see Fig. page 117) will be found necessary and economical.

Some farmers are in the practice of feeding their fowls from the hand, strewing it over the ground, while others throw down the corn in the ear in a heap, and permit the fowls to help themselves. This, however, is considered a slovenly and wasteful mode, and well calculated to invite rats and mice.

We have found it more economical to keep grain constantly before them, and for that purpose adopted feeding-hoppers, or fountains. Before adopting any plan, we examined the several works on poultry in our possession, but did not find anything to our liking; we then looked into Loudon's Encyclopædia of Agriculture, and there found the desired object, but too complicated and costly for our purpose; we however took a hint and constructed one ourselves.

The following is an *improved poultry-feeding fountain*, which we alluded to, published in the "Transactions of the Highland Society" in Scotland, and figured and described in Loudon's work on agriculture:

Fig. 21.

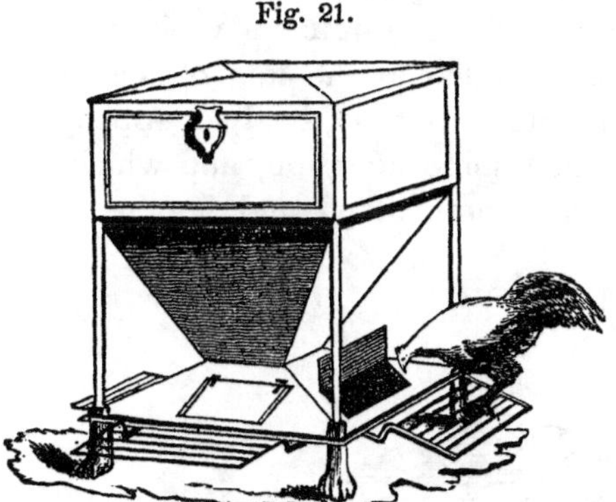

"It can be made to contain any quantity of grain required, and none wasted. When once filled it requires no more trouble, as the grain falls into the receiver below as the fowls pick it away; and the covers on that which are opened by the perches (the principles of which we do not understand), and the cover on the top, protect the grain from rain, so that the fowls always get it quite dry; and as nothing less than the weight of a hen on the perch can lift the cover on the lower receiver, rats and mice (which are very troublesome when grain is fed in the ordinary way) are excluded. It is astonishing, too, with what facility the fowls learn to leap upon the perches, and so open the cover of the receiver, which presents the grain to their view and within their reach." On their leaving the perch or platform, the door, either by a spring or weight, closes at once.

From the foregoing figure we had one constructed, of which the following figure is a fair representation, which we exhibited at the fair of the New York State Agricultural Society, held at Albany in 1842, which excited some attention, and which the committee highly commended and honored us with a diploma.

Fig. 22.

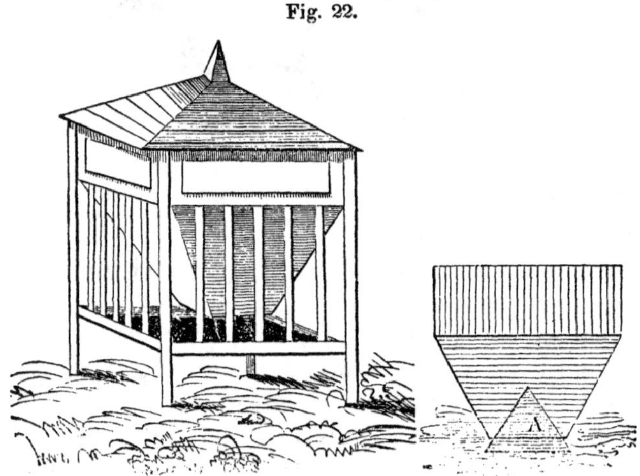

This feeding-hopper, as may be seen in fig. 22, is four square, two feet each way—posts eighteen inches long and two inches square. The upper section of the box is six inches deep, and the sides are morticed into or nailed to the posts. From the bottom of this square the slanting part or tunnel reaches to within half an inch of the floor, which should be six inches from the ground; the tunnel tapers from two to one foot; and in order to bring the grain within reach of the fowls, a cone (Fig. A is a section) is placed in the centre, as much smaller than the hopper as to leave half an inch space all around, which conducts the grain to the edge, where, as the fowls pick the grain away, more will fall, and keep a constant supply as long as any is left in the hopper. The slats on the sides prevent the fowls from getting in or crowding one another. This fountain will hold two bushels or more of grain, and protects it from wet and in a measure from rats. It occupies but little room, and from sixteen to twenty fowls can feed at the same time.

To protect the grain more effectually from rats and mice, we would suggest that the posts be made some two feet longer, and a platform of boards about one foot wide, placed round and fitted close up to the bottom, so that mice cannot climb up the posts and get in. This platform will be necessary for them to stand on when eating.

Before we had the foregoing made we constructed one on the following plan, which we had in use for four years, and it answers a very good purpose. It is very simple and easily constructed.

Fig. 23.

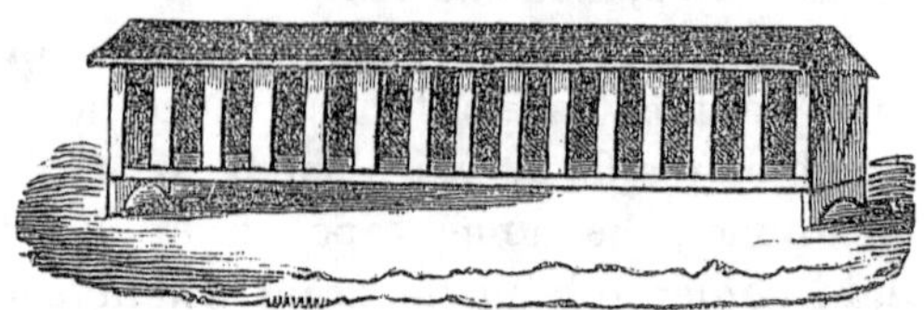

The dimensions are as follows; but it may be made of any desired length to suit the number of fowls kept. We made ours five feet long, nine inches wide; end pieces fourteen inches high, and the bottom raised five inches from the ground; the ends nailed to the bottom, and a strip of board four inches wide is firmly nailed on the sides, raised two or three inches above the bottom board, forming a manger or trough to prevent waste of food. Another strip three inches wide is nailed on the top in front to secure the ends. The hopper to contain the grain is formed of two pieces of board, nine inches wide, set between the ends forming a V, the upper edges lying against the front top strips and the bottom resting on some little blocks, sufficiently high from the bottom board to admit the grain to fall through as the fowls pick it away. It may be made to open and shut at bottom, to suit the different sizes of grain. The top or roof may be made of the same width as the box, or it may extend over the sides sufficient to protect the fowls from rain when feeding. Narrow strips of lath must be nailed to the top and bottom pieces, leaving space enough between them for the fowls to enter their heads when eating. It is open on both sides, and one of this length is sufficient for seventy-five or eighty fowls.

Mr. Ames says, "Hoppers can be made out of an old candle box, to be had at every grocery store for twenty-five cents. They are about eighteen inches long, twelve wide, and ten inches deep."

"Begin by taking off the lid and one of the sides, leaving the two ends, bottom, and one side remaining; then take the lid and cut a small strip off one of the ends so as it will slip in between the two ends of the box, placing the lower edge one and a half inches from the side and about an inch from the bottom; the other edge of the lid is to be brought out so as to reach the top and outside corners of the ends. In this position it will form a deep angular box

Fig. 24.

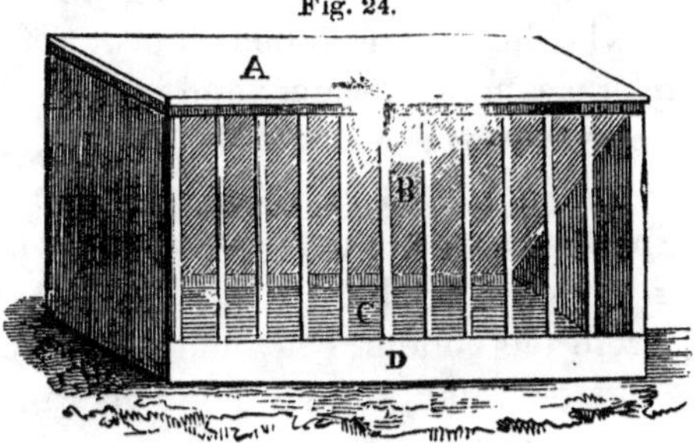

with a long aperture at the bottom. Two or three nails will secure it in this position. The lid now forming a slanting side, B, will be too wide and project beyond the ends; cut this strip off and nail it across the bottom of the hopper so as to form a trough, C, where the corn, when put in the angular box, will descend through the long aperture down into it. Take then the side which you have not yet used, and with a few tacks and some old shoe leather, make hinges and put on a lid, A. The front or open part of the hopper has a paling or row, D, of slats or wire about two inches and a half apart, so that the fowls can just get their heads between to pick out the corn. These wires should be brought out to the edge of the box, so that the fowls can but just reach the bottom of the angle or long aperture. Thus is a candle box turned into a food box or hopper."—*Rural Library.*

WATER FOUNTAIN.

There should, if possible, be running water in the yard, as fowls, like some other bipeds of larger growth, prefer clean, pure water; and in order to prevent their drinking by chance what is bad or corrupted, stone or wooden troughs, or what is much better, fountains similar to the following cut, which is nothing but a keg set on a stool, on end, with a small tube extending from the bottom to a shallow dish or pan, which should be small, so that the fowls cannot get into and soil the water. A jug, carboy or demijohn may be substituted, and on some accounts the glass vessel may be preferred, as it can be more readily perceived when empty.

Fig. 25.

When the hen is confined in a coop with her little family of chickens, they require considerable water, and if a vessel is put where she can get it, the chickens are very apt to get into it and not only soil it, but often get their down wet, which chills and much injures, if not kills them. To remedy this, we adopted the following fountain for her and her brood, which

Fig. 26.

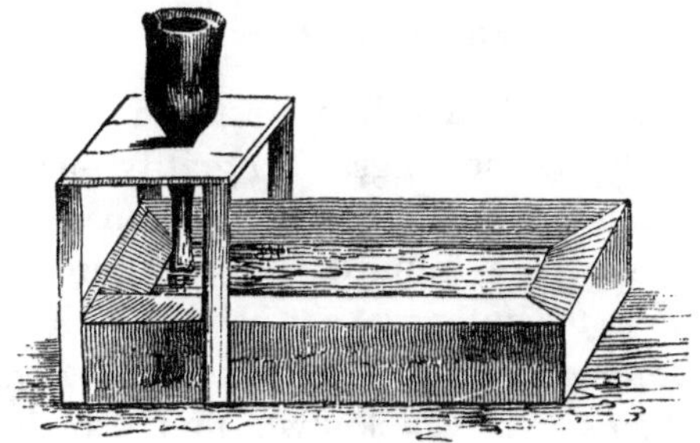

we found to answer admirably well. We took a piece of two inch plank, four inches wide and one foot long, scooping it out about one and a quarter inches deep, forming a shallow trough like the figure; nail four strips of lath on the sides of one end of the trough, and on the top at a proper height nail the laths to a square board the same width as the trough; cut a round hole in the top sufficiently large to receive the neck of a champaigne bottle, allowing the nozzle to reach within one-fourth of an inch of the bottom of the trough. When filled, the neck of the bottle is set into the hole, and the water will run out until the trough is nearly full or above the nozzle of the bottle, when it will stop, and remain so, until the water is displaced in the trough, which will keep about the same height until the bottle is emptied.

We also made a very good fountain for the same purpose by taking the bottom dish of a flower-pot, and made a frame around it similar to the frame at the end of the trough, and inserted a bottle in the same way.

PLANS FOR NESTS.

The hen is a prude, and likes to steal away in some sly place to deposit her eggs. To gratify their organ of secretiveness, the following ingenious plan for fixed nests we take from the "American Poultry Book," which, the author says, "has lately been contrived in Connecticut, and I have tried with complete success. Hens are well known to be anxious to deposit their eggs in secluded places. *The secret nests* here alluded to are well adapted to satisfy this propensity. They are made thus:

Fig. 27.

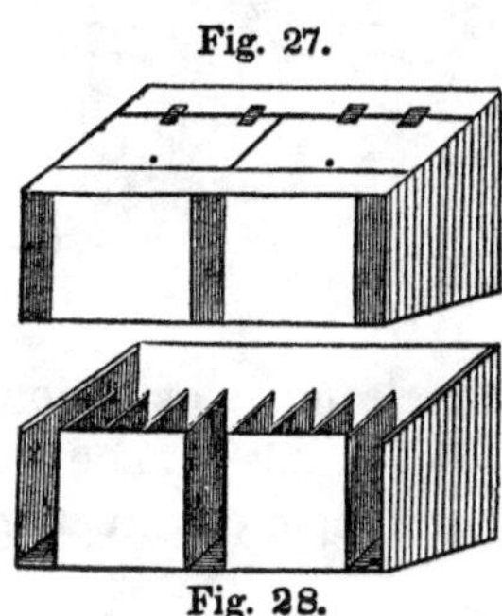

Fig. 28.

"Place a platform of boards two feet wide, and say ten feet long (though it may be made of any length, against a building or a close board fence, about three feet from the ground. Along the outer edge of this platform nail a board lengthwise and upright, about one foot high —leave a space open in the middle and at each end eight or nine inches wide, and divide the remaining space inside the nests a foot square; this leaves a passage way nearly a foot wide behind the nests. The top must slope from the wall, and open partly or entirely with hinges. These nests are easily examined, and give the fowls all the secrecy they seem to require. Fig. 27 shows the appearance of this series of nests when closed. Fig. 28 exhibits a view of the interior arrangement."

CHICKEN COOPS.

The most common method employed for the purpose of confining the hen with her young brood, is to drive stakes into the ground in front and make a pen about two feet square and cover with boards; but a

Fig. 29.

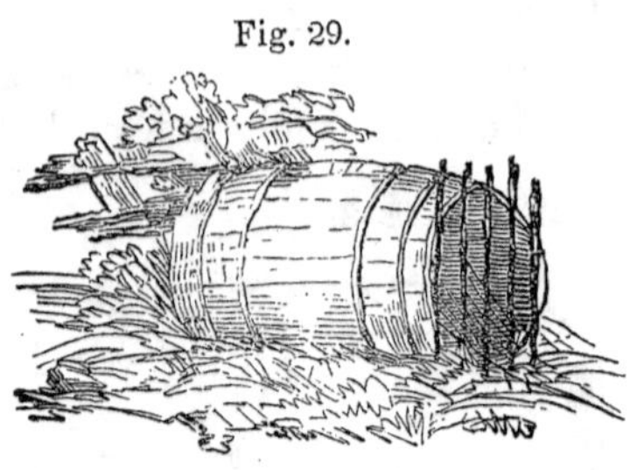

better plan is to lay a flour barrel on its side, with one end out, and drive a few sticks into the ground in front, This makes a very dry and comfortable coop, protecting them from rain and winds.

In England, hampers or baskets made of willow and open, inverted, are used; but in this climate the following plan will be found of simple construction and of trifling expense.

Fig 30.

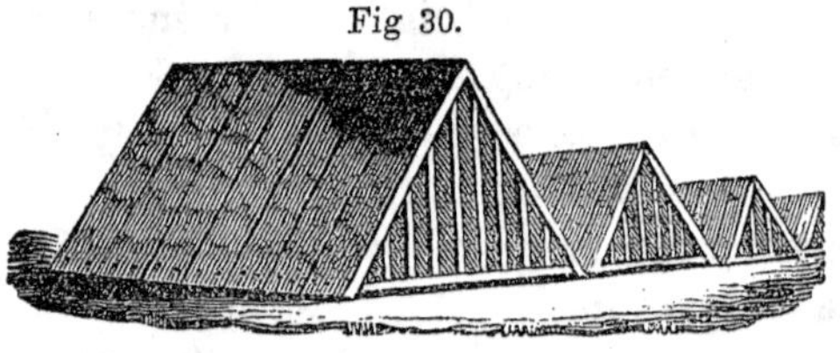

The above figure represents the marquee-coop, which we have used for several years, and find it answers a good purpose. It is formed by nailing boards two feet in length in such a way as to form two parts of a triangle, the ground forming the other, as in warm and dry weather we consider it best to have them next the earth; but early in the spring, when the weather is cold and the soil wet, a platform of boards or an old door should always be put under the coops. It should be at least two feet deep, or if three feet so much the better, and one end may be boarded up tight, and the other secured by nailing strips of lath, in the form of grates, leaving sufficient space between them for the free passage of the chicks without admitting the hen. In front there should be a broad piece of board of the same length as the front of the coop to feed them on. This board may be secured to the coop with leather hinges, so as to admit of its being raised up towards evening. This answers the double purpose of protecting the chicks against the smaller noxious vermin, such as rats, &c., during the night, and of preventing them from wandering about in the dew and wet grass in the morning. We have heard of keeping chickens under old forcing frames, and should think it an excellent plan where they are at hand.

Fig. 31.

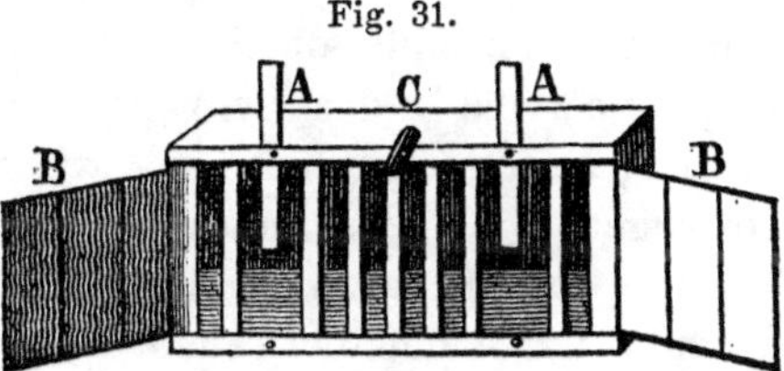

Mr. Lawrence Smith, a writer in the 8th vol. of the Cultivator, says: "The above cut is a coop of my invention, which I think is very convenient. It may be made of inch boards, long enough to admit of any number of fowls. A, A, slats raised for admitting the hens—B, B, doors to open and shut at night, to prevent the intrusion of any kind of vermin; C, button for fastening the door."

In all cases a warm, dry, and quiet place should be chosen for the coops, near the house, on account of the convenience of feeding them, and where the chickens are not in danger of being trod on either by man or beast, nor where the hen will be roasted by the hot sun, or where there is danger of the chicks being carried off by the hawks or crows. To make them thrive, sand or dust should be near at hand where they can wallow and bask themselves.

The coops should not be placed too near each other, as the chickens are apt to mix and get into the wrong coop, and some hens are so cross, that they often kill a strange chicken.

It has been our practice to place hens with their chickens in the walks of our garden, at least fifty feet apart; where they not only obtain their livelihood, but are of great service in destroying large numbers of bugs, worms, insects, and their eggs, which are so injurious to vegetation. We found some difficulty, however, in this, for the hawks would pounce upon them, and where the vegetables, such as beans and peas, were pretty rank, the rats will take shelter and catch the young chickens when they run among them. It is well to look to these evils, and we would also advise their removal after they are one month or so old, or they will become so attached to the garden that it will be difficult to keep them out.

At the end of six weeks the hen may be set at liberty after the dew is off in the morning and the weather fair, and if the moveable coop be employed, it may be propped up with a stick, and the hen will return to it of her own accord, at night, when it may be let down and kept so until the dew of the morning is dried off. At the end of two weeks more they may be turned into the poultry-yard.

As they will at first hardly receive fair play in the distribution of food, it will be necessary to prepare for them a feeding-coop, so that they may enjoy their food without being disturbed by the older fowls. This we effected by driving strips of board or stakes in the ground, leaving spaces between just wide enough to prevent the grown fowls from entering, encircling a space from four to five feet in diameter and about two feet high, covered with boards, through which was a small aperture or door, where the feed was put into the hoppers of the feeding box, which was made on purpose.

CHAPTER VII.

GALLINACEOUS FOWLS.

Their origin—not from the pheasant—wild varieties—Bankiva—Jungle—Fork-tail.

GALLINACEOUS FOWLS—ORIGIN, &c.

THE cock, by some writers, was supposed to be of Persian origin; but the period of their servitude is hidden in the remotest ages of the world. The acquisition of the fowl species has not, in all probability, been an easy conquest; to succeed in bringing them into complete bondage, a long series of attempts and cares has doubtless preceded the successes we now enjoy, without being acquainted to whom we are indebted for them. The species has been since propagated and introduced into general use throughout the whole world, from east to west, from the burning climate of India to the frozen zone. They may be looked upon as a blessing to humanity. Among every polished nation on earth, and even among nations half-civilized, but united in sedentary societies, there is no country habitation around which fowls, more or less numerous, are not met with, which man rears, shelters and nourishes, and which are called *cocks* and *hens*. They are a species which art has almost entirely wrested from nature; fowls are everywhere seen in a domestic state, and wild ones are scarcely to be found anywhere; it is not long since it is positively known where the latter still exist in small quantities.

Oliver de Serres says, "Among the moderns, *I am the first* that had seen fowls in a state of liberty. On my return from a first voyage to Guiana in 1795, I published a note on the subject of the wild cock and hen, which I have every reason to think natives of the hottest countries of the new continent. In travelling over the gloomy and inextricable forests of Guiana, when the dawn of day began to appear, amidst the immense woods of lofty trees which fall under the stroke of time only, I had often heard a crowing, similar to that of our cocks, but only weaker. The considerable distance which separated me from every inhabited place, could not allow one to think this crowing was produced by domesticated birds; and the natives of those parts, who were in company with me, assured me it was the voice of wild cocks. Every one of the colony of Cayenne, who have gone very far up the country, give the same account of the fact. Some have met with a few of these wild fowl, and I have seen one myself. They have the same forms, the fleshy comb on the head, the gait of our fowls, only they are smaller, being hardly larger than the common pigeon; their plumage is brown or rufos."

Some older travellers have spoken before of these wild fowl of South America. The Spaniard Acosta, provincial of the Jesuits at Peru, has positively said "that fowls existed there before the arrival of his countrymen, and that they were called in the language of the country, *talpa*, and their eggs, *ponto*. The ancient Mexicans had reduced these small fowls to domestication; they called them, as Gemell Carreri informs us, *chiacchialacca;* and he adds, that they were similar to our domesticated fowl, except that they had brownish feathers, and that they are rather smaller. A fresh testimony, that of a traveller who has been all over Dutch Guiana after me, is again come in support of facts already certain. Captain Steadman has observed that the natives rear a very small species of fowls, whose feathers are ruffled, and which seem to be natives of that country." (*Voyages to Surinam, and in the interior of Guiana.*) It is then an indisputable fact, that a tribe of wild fowl, very much like our cocks and hens, exists in the inland parts of South America. One cannot reasonably suppose that this tribe springs from birds of the same genus which Europeans have transported thither, since they are only met with very far from any inhabited place; that there is a remarkable difference in the size of these and the common fowl; and that, according to the assertion of Acosta, they existed in Peru before the arrival of the Spaniards.

But a learned traveller, to whom ornithology in particular is indebted for many capital discoveries, M. Sonneret, has again found the species of the wild fowl on the antique land of India, in the mountains of the Gautes, which separate Malabar from Coromandel. More successful than other travellers, M. Sonneret took home two birds, a male and a female, of the Indian tribe, and published a description of them in his *Travels to the Indies and China;* and he has taken them to be the primitive stock, whence had sprung all the tribes of our domestic fowl. He concurred in the opinion of Buffon, that most of our varieties of domestic fowl have proceeded from a single type; and that the differences which we perceive among them have resulted from accidents of climate, domestication, and crossing of varieties. Sonneret, who did not or would not know of any other species of wild cock than this—for he speaks slightly of the authority of Dampier, who mentions that he saw wild cocks in the Indian Archipelago—naturally enough concluded that in this jungle fowl he had found the primitive stock. Subsequent inquiries have, however, confirmed the statements of Dampier, not only as to the existence of species of wild fowl in the Indian Archipelago; but it is also admitted that the *Bankiva* species in *Java*, and the *Jago* species in Sumatra, more nearly approximate to our common fowl than that now under consideration, and to which Sonneret refers. Upon the whole, it seems that our varieties of domestic fowl proceed from mixtures of original species. Practical observers arrive at much the same conclusion on this point with scientific naturalists. It is thus, for instance, considered in India, that our game-cock originated from a mixture of the jungle cock with wild species in Malaga and Chittagong. Altogether, however, it must be admitted that, on this disputed point, very little is actually known; and the domestication of the bird ascends to such remote antiquity, that it seems hopeless to determine the era, and still more hopeless to ascertain the original species with precision.

"In most of the works on poultry," says a late author, "the domestic cock is declared to be a pheasant, and we accordingly find it described under the improper name of *Phasianus gallus*. This conveys an erroneous idea. According to modern naturalists, more especially a Dutch writer, *Temminck*, our domestic cock must be separated from the pheasant (*Phasianus*), and form a distinct genus under the name of *Gallus*, which had long since been proposed for it by a French naturalist. It is thus described."

GENUS GALLUS.—*Brisson.*

Bill smooth at the base, thick, slightly curved. Nostrils covered by an arched scale.—Generally erect, fleshy crest on the head. Throat, with fleshy wattles on each side of the lower mandible. Ears naked. Feet armed with strong spurs. Anterior toes united by a membrane as far as the first joint. Tail, 14 feathers, compressed, more or less arched, ascending. *In the female*, the comb and wattles less developed, and the tail wants the long pendent feathers.

"The original stock or species from which our common cock is derived is unknown. It is now, however, generally supposed to have sprung from a species (*Gallus Bankiva*), still abundant in a wild state in the jungles of Sumatra and Java. *Cuvier* supposes this to resemble most our domestic fowl, although *Temminck* thinks that the varieties *Morio*, Negro-fowl, *Lanatus*, Silky hen, *Crispus*, Frizzled hen, *Ecaudatus*, Rumpless hen, originated from other species, as yet undiscovered in their native state; as they all, however, breed freely together and produce prolific offspring, this may well be doubted. Hitherto, all the known species are natives of Asia, and it is very doubtful whether America contains any species from which our common cock can have been derived. There are, however, a few authors whose travels would seem to imply the contrary, but their testimony scarcely amounts to any degree of certainty. *Acosta*, the Jesuit, asserts that the common cock and hen existed in Peru before the arrival of the Spaniards. *Sonnini*, who travelled through the forests of Guiana, saw a small bird about the size of a pigeon, with a fleshy crest on its head, and its cry resembled exactly the notes of the domestic cock. *Steadman* also mentions that the natives in the interior of Dutch Guiana raise a small species of fowl which appears natural to the country."

VARIETIES—WILD.

I shall begin with a description of the wild species, natives of warm climates in the East, *supposed* to be the origin of our tame varieties.

Fig. 32.

BANKIVA FOWL.

The Bankiva fowl is a native of the Island of Java, and is characterized by a red indented comb, red wattles, and ashy-grey legs and feet.

The cock has a thin indented or scolloped comb, and wattles under the mouth, the tail a little elevated above the level of the rump, and the feathers somewhat disposed in the form of tiles. The feathers of the neck are long, falling down, and rounded at the tips, and are of the finest gold color. The head and neck are fawn-colored; the wing-covers are dusky brownish and black; the tail and belly are black. The hen is of a dusky ash-grey and yellowish color, and has her comb and beard much smaller than the cock, with no feathers on the neck besides the long hackles.

Specimens of these fowls, male and female, were brought from the Island of Java by M. Leschenaust, and deposited in the Museum of Paris. They inhabit the forests and borders of woods, and are exceedingly wild. On examining this species, it will be found to exhibit many points of resemblance with our common barn-door fowls of the smaller or middling size. The form and color are the same, the comb and wattles are similar, and the hen so much resembles the common hen, that it is difficult to distinguish it, except by the less erect slant of the tail. This rise of the tail is much more apparent in the male; but it may be observed, that in all the wild species known, the tail does not rise so high above the level of the rump, nor is it so abundantly provided with covering feathers, as in the common birds. Probably the superabundance of nourishment, and the assiduous care of man, have contributed to the greater development of all their organs. Different tame breeds, indeed, such as the tufted fowl, the Hamburgh, double-combed varieties, and others. show that domestication, probably on account of superabundant feeding, produces infinite varieties.

The Bankiva cocks very much resemble, both in

form and colors, the tame Turkish and Bantam breeds; but the tail differs in being nearly horizontal and vaulted in the Bankiva, while, in the others, it is more raised, and forms two upright planes meeting above, and diverging below. The feathers which fall from the neck over the top of the back are, as in our fowls, long, and with divided plumelets or beards, the feathers widening a little, and being rounded. The colors of the plumage are exceedingly brilliant. The head, the neck, and all the long feathers of the back, which hang over the rump, are of a shining flame-colored orange; the top of the back, the small and middle coverts of the wings, are of a fine maroon purple; the great coverts of the wings are black, tinged with iridescent green; the quill feathers of the wings are rust-red on the outer, and black on the inner edges; the breast, belly, thighs, and tail, are black, tinged with iridescent green; the comb, cheeks, throat, and wattles, are of a more or less vivid red; the legs and feet are grey, and furnished with strong spurs; the iris of the eye is yellow.

The hen is smaller than the cock, and her tail is also a little horizontal and vaulted; she has a small comb, and the wattles are very short; the space round the eye is naked, as well as the throat; and on this space are some small feathers, distinct from each other, through which the red skin can be seen; the breast and belly are light bay, or fawn yellow; on each feather is a small clear ray, along the side of the mid-rib or stern ; the feathers of the base of the neck are long, with disunited beards or plumelets, of a black color in the middle, and fringed with ochre yellow; the back, the coverts, the wings, the rump, and the tail, are earthy grey, marked with numerous black zig-zags; the large feathers of the wings are ashy-grey.

The reasons for believing that the Bankiva fowl is the wild stock from which our tame varieties derive at least their main origin are, the close resemblance of their females to our tame hens, the nature of the feathers, and the forms and distribution of the barbs, which are absolutely the same in our tame cocks; and because it is in this species alone that the females are provided with a comb and small wattles, characters not found in any other of the wild species which are known.

As another wild species, the jungle cock, however, has been maintained by Buffon and others to be the origin of our barn-door fowls, it may be necessary to show by description that it is not so.—*Dickson.*

JUNGLE FOWL.

This species, which is three feet four inches in length, inhabits the great forests of India, continues to reproduce there in the wild state, and is clearly distinct from the domestic races reared by the Hindoos; as these resemble, in all respects, the other tame breeds of fowls in every quarter of the globe. M. Sonn ret, however, thought very differently, and prided himself much on the discovery, rejecting the statements of Dampier and others as to the existence of wild fowls in Timor, and other islands of the Indian seas.

The jungle cock is one-third less than our dunghill cock, and from the level of the feet to the summit of the head, the comb not being included, the wattles resemble those of our dunghill cock; but the naked parts of the head and throat are more considerable. The feathers of the head and neck are longest on the lower parts, and their form and structure are different from those of the same parts in other cocks, whether wild or tame. The mid-rib, or stem, is gross, and considerably expanded, forming a white stripe along the whole feather, as far as the tip, where it becomes broader, and gives rise to a sort of grisly plate of a rounded form, whitish, thin, and very polished. This grisly substance is found still more obvious upon the wings. The tip, indeed, of all the wing feathers forms a broad grisly plate, whose substance is solid, and very thick to the touch, though as shining and polished as in the feathers of the neck. The plates are of a deep red, and form, by their union, a plate of red maroon, which looks as if it were varnished. The jungle hen is smaller than the cock, has neither comb nor wattles, and the throat is covered with feathers—a very marked distinction from our domestic hens. The space around the eye is naked, and of a reddish color. All the plumage of the under parts is similar to that of the cock, only more dull; she has not the long feathers at the base of the neck, nor the thin gristly plate at the tips of the feathers. All the parts are more or less blackish-grey, with white rays occupying the middle of the feather, and following the direction of the stem.

The jungle fowl, in a word, seems to be as different from any known variety of our tame fowls, as a hare is from a rabbit, or a goat from a sheep, and the fact, that a jungle fowl is not domesticated in its native country of India, while our dunghill fowls are common there, seems to settle this question beyond appeal.—*Dickson.*

The cry of the jungle fowl is in some measure different from that of our domestic species; but there is much resemblance in their habits and disposition. The following lively statement on this subject is from "*Excursions in India*," by Captain Thomas Skinner, published in 1832.

"In some parts of the forests we saw several jungle fowl; they have exactly the same habits as the domestic poultry. The cock struts at the head of his hens, and keeps a strict watch over their safety. Whenever they were disturbed by our attempts upon them, he flew to the highest branch of some tree beyond our reach, and crowed with all his might, while his dames ran into holes and corners to escape our attacks; they are so cunning, that we found it impossible to get within shot of them with all the caution we could use. While intent upon capturing at least one, as we were creeping after them upon our breasts, lying occasionally like riflemen under cover of the unevenness of the ground to catch them, *en passant*, we came suddenly upon an ambuscade that very soon put an end to our sport.

"We were about midway up the face of a hill that was thickly covered with trees, and much clogged by shrubs and creepers that wound in all directions. On reaching the foot of the enemy's position, still advancing upon our breasts, and bending a keen eye upon the birds strutting before us, up rose, with a growl that denoted an offended spirit (for we had literally touched his tail), a large black bear; and, turning round, looked us in the face with the most undisguised astonishment. It was the most unsought, as well as most unpromising introduction I had ever met with. There was no time for parley, and getting upon our legs, we at once stood up on the defensive. This sudden metamorphose completed his surprise, and yelling louder than before, he set off as fast as he could shuffle from the extraordinary animals that had so unaccountably sprung up before him. We determined that discretion was the better part of valor, and began to retrace our steps, leaving the jungle fowl to benefit by the interruption."

It is seen by the foregoing description of the wild cock and hen of India, that the most striking dissimilarity consists in the wild fowls having no comb on the head, nor fleshy wattles hanging beneath the throat; but this difference is not sufficient to make this tribe be considered as other than that of the common fowl, in which, as it is known, a very ancient subjugation, removals and multiplications in opposite climates, differences of food, have produced numberless varieties, which, from all appearance, came originally from the wild fowl of the Gautes. There grow besides, among common fowls, and chiefly in the tribe of tufted fowls, individuals whose head is without a comb, and the bill beneath without appendages. It is equally very likely that the wild fowls of some desert southern parts of America are only the same tribe lessened and altered by the influence of the climate. But it will be said, how can heavy birds, which can hardly fly, be found on both continents? It might be possible to launch into dissertations of some length on this question. I shall abstain from it, and when the fact is certain, it ap pears to me to be of little use in searching in this place how it could exist? Do not other genera of heavy birds give instances of this habitation common to both continents? The quail is found in our own countries, and in those of America which border on the equator. There are in the same part many species of pheasants; and the ostrich, which never flies, is found again, with some restrictions, in Peru, Chili, and the land of Magellan.—*Main.*

The reasons for believing the Bankiva fowl is the wild stock from which our tame varieties derive at least their main origin are, the hens, the nature of the feathers, and the form and distribution of the barbs, which are absolutely the same in our tame cocks; and because it is in this species alone that the females are provided with a comb, and small wattles, characters not found in any other of the wild species which are known.

THE FORK-TAIL FOWL.

This curious fowl was first described by M. Temminck, in 1818. It is nearly two feet in length to the extremity of the tail. The cheeks are bare, the head furnished with a simple entire comb, and the throat with a single large wattle, springing from the centre; they are all bright red. The head, neck, and upper parts of the back, are remarkable from the short and rounded form of the feathers—of a dark metallic blue. The hanging feathers are of a rich metallic green, tinged with steel-blue. The bill, legs, and feet, yellow. The hen has a circle round the eyes only, naked, and of a livid tint. This bird is said to be very abundant in Java, and may be often seen during the day upon the edges of the woods and jungles, but possesses the same disposition of its cogenus and pheasants, and, upon the least alarm, runs to cover. They are not domestic, but they occasionally breed with the tame varieties—a curious fact, and showing the uncertainty with which the true origin is clouded.

CHAPTER VIII.

DOMESTIC VARIETIES.

The Java—Dunghill—Malay—Dorking—Spanish— Bolton Greys or Creole—Creeper, or Dwarf—Black Poland—Golden Top-knot —Negro—Russia—Ostrich—Turkish—Game—Silver Top-knot—Frizzled—Silky —Bantam— Bucks County—Rumpkin—Cochin China.

THE JAVA FOWL.

This is a singular breed of fowls, which partakes of the common fowl and the India fowl, peculiar to the island of Java, where they are seldom reared but for fighting; and are said to be so furious, that they sometimes fight together till the death of one or the other separates them. According to Willoughby, it carries its tail nearly like the turkey. The Sieur Feurnier informs us, that one of this species was kept in Paris: it has, according to him, neither comb nor wattles; the head is smooth, like that of a pheasant. This fowl is very high on its legs; its tail is long and pointed, and the feathers of unequal length, and, in general, the color of the feathers is auburn, like the vulture. It is generally supposed the English game-cock originated in, or is a cross of, this variety.

DUNGHILL FOWL.

This variety is a mongrel, though a common and useful fowl, and generally proves profitable, at least in this country; it is therefore that which, in general, is adopted.

The distinguishing characters of the dunghill cock are, a thin indented or scolloped comb, with wattles on each side, under the bill; the tail rising in an arch above the level of the rump; the feathers of the neck long and line-like. Their plumage exhibits endless varieties, which also differ among themselves, which probably arises from crossing with all sorts and varieties. The best are of a middle size, with dark or speckled colors, and black or slate-colored legs. Light-colored or white fowls are considered of tender constitution, and are not generally so good layers. Experience has taught us that common fowls with black legs are preferable for the product. Yellow-legged fowls look best when dressed, and are generally selected in preference; but their flesh is coarse, and not so savory and high-flavored as the darker-colored ones.

The influence of domestication in the domestic fowl having produced alterations in the entire forms of the body, and of the comb and wattles, it is difficult to indicate the races which owe their origin to the same stock or races; but those races most approaching the primitive species, have undergone the fewest changes from taming, and have produced, by inter-alliance, the breed of dunghill fowls with comb and wattles; or the breed of tufted fowls, with small comb and wattles, in which those juices, proper to form the comb and wattles, appear to have been expended in the production of the tuft of feathers which ornament the head; all those breeds of fowls whose legs and feet are more or less covered with feathers, the origin of which may be attributed to the superabundance of nutriment, that has in this instance produced feathers on the legs, as it has formed a tuft on the head of the tufted fowl; the race of Hamburg fowls, which have the head hooded, and the feathers stretched back over the ears.

MALAY FOWL.

This, if not the largest, is among the largest, of the gallinaceous tribe. In color, they are generally brown, or a dull yellow. They are what is termed serpent-headed, on account of having little or no comb and wattles. Their flesh is coarse, flabby, and of a dark color, and not fit for the table until full grown, when they make a very good substitute for a turkey. Mowbray says, "they are good layers, and being well fed, produce the largest of hen's eggs, and of the most substantial nutriment. Being too long legged, they are not generally steady sitters. They are large birds, coarse meat, and not worth breeding from."

"The Malay," says a writer in the American Agriculturist, "is an awkward, bony, leggy, cowardly race; wandering about for the first six months of its life with scarcely a feather to cover its naked-

Fig. 33.

MALAY FOWL.

ness, and almost for ever in coming to maturity, a wretched layer, and worse sitter, usually breaking half its eggs in the operation; an indifferent nurse, and never yielding in either its eggs, flesh, or appearance, half enough to compensate for the anxious and vexatious labor of its rearing. When half grown, or in moulting time, it looks more like a sand-hill crane than a domestic fowl, and although it sometimes gains a weight of seven or even ten pounds, its flesh is coarse, and lacks the delicacy and richness of the well-bred chicken. Its color runs through all shades, from a light yellow to a brownish black, with little variation on the same lines. They are withal, great eaters; and although at the south they may both thrive and lay better than at the north, they are not to be recommended as a valuable kind of fowl. Their eggs are large, of a buff, or light brownish color, sometimes almost speckled like the turkey's. They seldom lay more than ten or twelve at a litter. The outer shell is ofttimes very thin, and the under skin so tough and unyielding, as, in numerous instances, to strangle the chicken in its birth, requiring assistance to release it. On the whole, I have found this a most unsatisfactory bird, and although they have been praised for their great size for the table, and as a tolerable substitute for a turkey, no competent judge of the latter flesh would ever mistake the stringy, juiceless meat of the one, for the rich, delicate, and variegated flesh of the other; and what good liver would provide the one, when the other, much better, could be procured at half the expense? Crossed, however, in a small degree, with the common dunghill fowl, they give increased size both to their egg and body, and may, in mild climates, be of some value for that object."

Now hear what Dr. Kittridge says; "The Malay is a large noble fowl, weighing from eight to twelve lbs. They are good layers, eggs very large, and hatch well. They are hardy; I have never lost a chicken; come early to maturity, and their meat is excellent. I should think them superior to any other breed for market."

They have been made to weigh from nine to ten

pounds, and a cock of this variety is said to have stood on the ground and picked corn out of a flour barrel as it stood on end. A flock of these birds make quite an imposing appearance in the poultry-yard, and are quite too much for the hawks. Our portraits were taken from Mowbray's work on poultry.

Mr. Coons, a farmer in Rensselaer county, who is very curious in the breed of his fowls, commenced about twenty years ago, with the pure Canton or Malay breed, but finding them rather poor layers and unprofitable sitters, he first crossed them with the Dominique cock, and then introduced white cocks with yellow legs; and afterwards by selection, keeping in view the yellow legs, small comb and gills, and the small tail and form peculiar to the breed, has succeeded in retaining these characteristics, and reduced them to the size of our common fowls, and of a pure white color. Mr. Coons informed me that they were now good layers, good sitters, and notwithstanding he has been breeding "in-and-in," for the last ten years, but always selecting the best, I must confess, I have never seen a more healthy looking yard of fowls. There were about sixty in the yard, and my attention was first arrested by their pure white color, yellow bills and legs, small tails of the hens, and short plumes of the cocks.

It is a fact, worthy of remark, that one of the cocks has four spurs, which Mr. C. said broke out occasionally, and he could trace it back to the Dominique cock, which he first used twenty years ago.

Fig. 34.

DORKING FOWL.

This is a valuable and favorite variety in England, and takes its name from a town in the county of Surrey, where the breed is supposed to have originated, and where, and in its vicinity, they are still said to be found in great plenty and perfection. They have been but little known and scarce in this country until within a few years. Doctor E. Wight, of Boston, imported some in 1839. A. B. Allen, of Buffalo, F. Rotch, of Otsego, and a Mr. Chapman, of New York, imported some in 1841–3. Individuals, also, of this breed have occasionally been brought by vessels, who exchange their old stock in port with the dealers, for a fresh supply.

" In size," Mowbray says, " they rank, in the third degree, in the largest of our fowls; well shaped, having a long capacious body and shortish legs, and should have five claws on each foot. This is a distinctive mark, but of no advantage, but probably tracing their origin to the Poland; as it is said a Poland cock with a common white hen will occasionally produce a similar bird. The absence of a fifth claw is not, however, considered a proof of a spurious breed."

There can be no doubt that the production of two hind toes instead of one, is merely accidental, like that of two thumbs on each hand, sometimes observed to run in particular families; " but this," says Dickson, "is certainly not peculiar to the fowls bred about Dorking, in Surrey, for five-toed fowls are mentioned by Aristotle in Greece; by Pliny and Columella in Rome; and by Aldrovand in Italy, hundreds of years ago; the breed then, as now, being reputed good layers."

The writer of the article, " Poultry," in *Rees' Cyclopedia*, seemingly on the best information, says, " it is quite a mistake to suppose that the Dorking breed of fowls have uniformly five toes; but they are of a larger size than the ordinary dunghill fowls; the body long, and the eggs large. The flesh is rather yellow than white, and requires to be improved by castration. The colors are as variable as the dung-hill fowl."

Mowbray says, " the genuine color is of an ivory

white; the flesh is good-flavored, of a yellow or ivory shade, and highly esteemed."

"The most valuable variety for the table at present," says Main, "is the *Dorking* breed. This is pure white; and highly esteemed for whiteness and delicacy of flesh when served at table; they also fetch a high price at market. Among breeders, real Dorking cocks sell for from five to ten shillings (sterling) each."

"This breed makes an excellent stock for the farm or market. They fat well, lay well, and rear well; are handsome alive, and show delicately white and advantageous when plucked and dressed for market. Their feathers also, being fine and of a good color, can be substituted many of them for geese feathers; consequently they will bring a higher price."

From the specimens we have seen, we have no reason to believe that *color* is any criterion of purity. The first we ever saw were a pair presented to us by Dr. Wight of Boston, which he imported. The hen was of an ivory white, and the cock was a lead or hawk-colored fowl. Since then, we have seen some bred by Mr. Allen, of Blackrock, and Mr. Chapman, of New York, which were of various colors, but generally speckled. Our portraits were sketched by Mr. Rotch, from specimens in his yard, one of which was imported.

A gentleman in Boston, who has paid considerable attention to the rearing of poultry, says in a letter to the author, "so far as my experience has gone, the Dorkings are *decidedly* the best breed for laying; the eggs come abundantly, and are of the largest size, except when they have been bred 'in-and-in,' too much. (See Walker on "Intermarriage," article on fowls.) I have already seen the effect, and therefore hope to receive a new lot of Dorkings during the summer." Mr. C. gives it as his opinion also, "that the Dorking is the best for laying, as well as for making good-sized poultry—though many prefer the Poland to all others as steady layers." After six months more experience, the same gentleman says, "In regard to the *Dorkings*, I am still strongly prepossessed in their favor; as layers, they are certainly very prolific; as an instance, one of my neighbors had a pullet which was hatched in May; in the *same year*, the pullet began her litter of eggs and hatched out her chickens, before the first of December ensuing. This is only one of many instances which could be advanced in their favor."

Another breeder says, "The Dorkings sustain a high character in England, as layers, but mine have not laid better than our common hen. Their meat is fine, and they are very hardy, and might be raised for market. They weigh from six to nine pounds."

Another writer says, "The Dorking fowls stand first in the estimation of those who have raised them. They will weigh from five to eight pounds. Their bodies are large, and better proportioned than any others, being long, full, and well-fleshed in the breast; have short legs, and beautiful plumage, with five, instead of four toes; are good layers, good sitters, and good nurses. Their eggs are large, clear, white, and of excellent quality. When caponed, they weigh from ten to twelve pounds."

L. F. Allen says, in the American Agriculturist, "the Dorking is a fine large bird, weighing, when at maturity, five to eight pounds. They are large bodied, and of better proportions, according to their size, than any breed I have yet seen; their bodies being very long, full, and well-fleshed in the breast and other valuable parts. They are short-legged, thickly feathered, with fine delicate heads, both double and single combs, and a shining, beautiful plumage. The color of their legs is white, or flesh-colored, having five, instead of four toes, the fifth being apparently superfluous, and rising like a spur from the same root as the heel toe in the common varieties. This is a distinguishing mark of the variety. They are most excellent layers, good and steady sitters, and kind, careful nurses. Their color is various—from nearly white to almost black, many of them beautifully variegated. They are the capon fowl of England, and are bred in great quantities for the luxurious tables of the wealthy classes in the counties about London. In America, they are a scarce bird. I never saw one till the fall of 1841, when a friend, by whom I sent, brought me out half-a-dozen from England; and although they were but chickens when they arrived, and, from their long confinement on the voyage, miserably poor, and full of vermin, they rapidly improved, commenced laying during the winter, and have thus far exceeded any other fowls I ever kept, in their good qualities. The young have proved very hardy, and easy to rear. The males, of which I imported two, are large strong birds, and the hens are all I could desire of them. Their eggs are of a large size, clear, white, and excellent in quality. For capons, they no doubt exceed all other fowls whatever, often weighing, full grown, ten or twelve pounds. This variety I have determined to keep for my own purposes."

In the same paper, under the same signature, we find the following:—" I have had frequent inquiries from persons at a distance, for pairs of the Dorking fowls you brought me last year from England. As I ordered them for my own use, and not for sale, I take pleasure in giving them away as they can be spared, to any of my friends who feel the least desirous of obtaining them. In those raised this year, I have been unfortunate in the sexes; having at least three males to one female. I have now ten pullets, and several cocks, and very beautiful they are too. The hens are pheasant-shaped; with a clear and beautiful head and throat; and a deep heavy brisket, like a Durham heifer. The cocks are magnificent—variegated in color, with a brilliancy of plumage I never saw surpassed, and rarely equalled. I had the curiosity the other evening to weigh a pair of them. The cock, now seventeen months old, weighed eight pounds, and the hen of the same age, five pounds, both in common flesh only. I have no doubt, if fattened, the one would have weighed ten and the other at least six pounds."

"A friend, who is a great fancier of fine animals, and who possesses in his various breeds of horses. cattle, sheep, swine and dogs, &c., as fine specimens as the country can produce, in a note to a letter recently received, observes, 'The Dorking cock you sent me is a superb fowl; I shall hereafter make this my chief stock of hens. Cannot you send me a pullet in the spring?'

"Although in the depth of winter, with over a foot of snow on the ground, the hens lay daily, running out in the severest cold."

SPANISH FOWL.

A large fowl, the plumage black, flesh white and delicate, but inferior in size to the old Duke of Leeds' breed. They are well adapted for *Capons*, and produce eggs nearly equal in size to those of the Malay hen.

Dickson says, this is evidently a variety of the Poland or crested fowl, distinguished by the thighs and belly being of a velvety black color; its iris is yellow, and its eyes are encircled with a ring of brown feathers, from which rises a black tuft that covers the ears. There are other feathers nearly like those behind the comb and beneath the wattles, and broad, round black spots on the breast. The legs and feet are of a lead color, except the sole of the foot, which is yellowish.

The Spanish breed lays very large eggs, like the Paduan, and is esteemed for fine flesh. Our portraits were taken from Mowbray.

Fig. 35.

THE SPANISH FOWL.

BOLTON GREYS, OR CREOLE FOWL.

"This variety," says Mowbray, "apparently the crack breed of their vicinity, but entirely unknown in the metropolis (London), is described by the Rev. Mr. Ashworth, near Bolton, Lancashire, as follows: 'small-sized, short in the leg and plump in the make. The color of the genuine kind invariably pure white in the whole cappel of the neck; the body white, thickly spotted with bright black, sometimes running into a grizzle, with one or more black bars at the extremity of the tail; they are chiefly esteemed as very constant layers, though their color would mark them for good table food."

From the following description of a breed of

fowls known in Pennsylvania, as the Creole breed, we conclude they are the "Bolton Greys," under a new name. In a letter from Thomas P. Hunt, Esq., to the Editor of the New England Farmer, he says, "I have not been able to ascertain the '*habitat*' of the Creole. They are white, with black spots all over, except the neck, which is perfectly white. Their tails are more fan-like, or displayed, during laying time, and their rumps present a fuller or more elevated appearance than other fowls. The ends of the tail feathers are generally blackish. They are capital layers—poor sitters."

Doctor Rufus Kittridge, of Portsmouth, N. H., who seems quite a connoisseur in the poultry business, says, in a letter to the author, "The first account I ever saw of the Creole fowl (which led me to procure them), was in the New England Farmer, Vol. xviii. No. 39 and 43.

"The Creole is a small fowl, about the size of our common hen; the greatest layers I ever saw. I received my two from Philadelphia on the 26th of last April; on the 5th of May, they both laid. In twenty-two days I had forty-one eggs, and in fifty-four days, I had ninety-five eggs. They are never inclined to sit. Their color is a white ground with black spots on their bodies—their neck white. The end of the tails of both cock and hen is black; the cocks are mostly white; the hen, when laying, spreads her tail like a fan. They are very broad on their rumps. One of my hens has yellow legs, the other slate-col-

Fig. 36.

BOLTON GREYS, OR CREOLE FOWL.

ored. They are a very hardy fowl, and I value them the most of any I have."

A friend of the author, who resides on Staten Island, obtained some of this breed, under the name of "Leghorn Fowls." They answer the foregoing description, and the portraits, with the exception of their combs, which are high, thin, toothed, and fall over on one side. They are a little under the ordinary size, and esteemed good layers, and make a pretty show in the yard.

CREEPER, OR DWARF FOWL.

This variety is described by some authors as not larger than a pigeon, and differs from the Bantam chiefly in size and in the shortness of its legs.

The Acoho is described as very small, with a circle of feathers about the legs, a thick tail, which it carries straight, and the ends of the wings black. Other varieties, said to come from Cambodia and now found in the Philippine Isles, have the legs so short as to drag the wings on the ground.

In addition to these Buffon mentions a sort of fowl in Brittany, which are always obliged to leap, the legs being so short. They are the size of a dunghill fowl, and kept as being very fruitful. The hens will hatch thirty eggs at a time. Some think these dwarf fowls are the Hadrian breed mentioned by Pliny. Aldrovand, two thousand years ago, describes the dwarf hen as all black, except the quill-feathers of the wings, which are white at the ends, with some crescent-like spots on the neck, and a yellow spot around the eye. The head is furnished with a crest of feathers; the comb small and dark colored; the feet yellowish; the claws equal and very white. He does not mention the cock.

The Creeper is found occasionally in our farm-yards, and are considered less troublesome in gardens. The shortness of the legs prevents too free use of them on new made beds. In size they are generally below the common fowl of the country.

Fig. 37.

BLACK POLAND TOP-KNOT FOWL.

These, like the brave people from which they derive their name, are every way commendable, and are recommended to the "chicken fancy."

The Poland Fowls, as they are generally called, were, according to English authors, said to be imported from Holland. Their color is a shining black, with a white top-knot of feathers on the heads of both cock and hen. The head is flat and surmounted by a fleshy protuberance, out of which spring the crown of feathers or top-knot, white or black, with the fleshy King David's crown, consisting of four or five spikes. They are not so thickly covered with feathers as some other breeds, and still less so with down. The true breed is rather above the middling size; their form is plump and deep, and the legs of the best sorts are not too long, and most have five claws. The top-knot of upright, white feathers, covers so much of the head as almost to blind the eyes; indeed some require clipping, or they would become an easy prey to the hawks. The contrast of this perfectly white crest with the black plumage, is truly beautiful; but the top-knot of the cock differs from his hen, hers being broad and erect feathers, while his are narrow and hanging down in every direction, but they must be perfectly white and the rest of the plumage perfectly black; broken colors, it is said by some, show a cross breed.

Mowbray says, "The Polanders are not only kept as ornamental, but they are of the most useful varieties, particularly on account of the abundance of the eggs they lay, being least inclined to sit of any other breed, whence they are sometimes called *everlasting layers*, and it is usual to set their eggs under other hens. They fatten as quickly as any other breed, and in quality similar to the Dorking; their flesh perhaps more juicy and of a richer flavor." They are a quiet, domestic fowl, neither quarrelsome nor mischievous, and their eggs of a good size, fine flavored and thin shells. Mowbray states that his five Poland hens in eleven months, laid five hundred and three eggs, weighing, on an average, one ounce and five drams, exclusive of the shells, making a total weight of $50\frac{1}{4}$ pounds.

"Besides the Polanders," says Mowbray, "there is a small variety imported from Holland, called *every day hens*, which are everlasting layers. Their eggs generally are not so large as those of the common hens, nor equally substantial and nutritious."

"The whole breed of crested fowls," says Boswell, "is much esteemed by the curious, and reared with care."

Those who desire to propagate any singular varieties, must keep them apart and not allow them to intermingle with those of a different color. They are generally esteemed in proportion to the beauty and rareness of their tints. Such are the gold and silver ones, the white and black ones, the black top white ones, the *Das Hauben Huhn* of Dr. Bechstein—which we have never seen, but imagine they must be very splendid when pure. We have in our yard a pullet, of a greyish colored body, with a black neck and crest.

Some travellers assert that the Mexican poultry are crested; but these, as well as all the rest on the continent of America, have been introduced from the ancient continent. They are equally abundant at the Cape of Good Hope, where their legs are feathered. In Egypt they are very much esteemed on account of the excellence of the flesh, and are said to be so common as to be sold for two pence or three pence each.

There can be little doubt that all the fowls with crests, have originated from intercrossing with the Paduan or Polish.—*Buffon.*

We have in this section, fowls entirely white with equally large and full top-knots, purely white, and they make a very fine appearance in the poultry-yard, in contrast with the black ones.

"The Poland is," says L. F. Allen, in the American Agriculturist, "a shining black in color, with a beautiful white tuft on its head, a medium size, a good layer, seldom sitting to hatch, rather tender to rear while a chicken, and more thinly feathered and not so hardy in colds and storms as the common hen. In a great part of the United States it will thrive successfully, and lay as many eggs as any other fowl, perhaps more. Its flesh is good. On the whole, a handsome and profitable fowl. There is a white variety, without a feather of any other color. These are very beautiful, but not quite so hardy as the black. There is also a splendid gold and black, or pheasant colored variety. These are scarce in the United States. I have seen several beautiful specimens imported from England; but was never able to obtain any for breeding. These colors are more propagated by the poultry-fanciers than others, and are seldom to be had of them."

Mr. Giles, of Providence, who has paid considerable attention to the subject of poultry, in a letter to the author says, "If *eggs* is the only object in view, then as far as my experience goes, the Poland fowls are the best layers, seldom if ever wanting to sit."

From the foregoing it seems to be generally admitted that the Poland fowls are the best breed for the production of eggs, but the author has found, in his experience, a cross of the hawk-colored or Dominique fowl, are not only good layers, but good sitters, and the best of mothers. They generally have light-colored legs, and always look young and handsome in market.

The portraits of Poland fowls at the head of this article, were taken from living specimens in the possession of the author, by Van Zandt.

GOLDEN TOP-KNOTS.

This is an ornamental variety we esteem above all others for their splendid plumage of bright and odd contrasted colors. Our portraits were taken from specimens in our possession, from which we hope to perpetuate the breed. In size they are less than the Polands, but of beautiful symmetry; bodies short, plump and compact; tails stand high, and are long and large in proportion to the body. Their color is a yellow or buff-colored ground, with small round black spots; crests or top-knots large, and of mixed colors. Some of the cocks are red with black breasts; wings tinged or spangled with bright yellow or gold color.

They are very scarce, and it is supposed they originated from a cross of the black Poland cock and a common yellow hen, and purely accidental. We have no knowledge of their origin; they have been

Fig. 38.

GOLDEN TOP-KNOTS.

bred only by one person in the vicinity of Albany, and by always selecting those of the most odd and fanciful colors for breeding, produced a breed as strongly marked in character as the Dorking or black Poland.

Boswell says, " There is an ornamental sub-variety known as the *Golden Poland*, with yellow and black plumage."

Buffon, in speaking of the crested cock, says, " The breed of crested fowls is that which the curious have most cultivated, and what generally happens when things are closely examined, they have observed a great number of differences, particularly in the colors of their plumage, which serve to distinguish a multitude of races, that are the more esteemed in proportion to the beauty and rareness of their tints, such as the gold and silver ones," &c.

They are good layers, though their eggs are small, but rich in quality; flesh juicy and delicate. When young they do not hatch well. They make a splendid appearance in the poultry-yard, and are much admired. They are rather tender in constitution, and it is difficult to raise their chickens.

NEGRO FOWL.

This bird is a native of Africa, and differs from all others, as it has the crest, the wattles, skin, bones, and feathers, almost always of a black color; the flesh is white and good. It is common in Java, the Philippines, and in some southern parts of Asia as well as Africa. Those that have been carried to Europe, are only kept for curiosity.—*Dick.*

"In France," says Main, "it is reared merely to gratify the curious; for when dressed, its flesh turns black and is ill-tasted; it seems as if it were boiled in ink." How doctors will disagree!

From this mixture of Negro fowls with the other breeds, arise mongrels, which usually retain the black crest and wattle.

This breed of Negro fowls has been carried to and increased in the hot parts of America. "In Paraguay," says M. D'Azara, "Buenos Ayres, and in the Cordilleras of the Andes, there are tame fowls of common and other breeds, which do not differ as to the shapes, and which have their feathers, legs, comb, wattles and skin, black as the negroes of Guinea. When dressed, their skin is still black; their flesh is more insipid and of a darker color than that of the common fowl, and their bones are plainly more opaque. Their eggs are white, and some people value these fowls because they are said to be more fruitful, and their flesh accounted more fit to be given to sick persons. They probably descend from the common Spanish or Canary breeds."

RUSSIAN FOWL.

A few of this very singular and unique variety of fowls, was imported in 1842 from Moscow, by Dr. Wight, of Boston, from which our portraits were taken. In a letter accompanying the portraits, the Doctor says, "I herewith send you a rough sketch of a cock and hen of the Russian or Siberian Fowls. They came to hand a few weeks since, and are perfectly described in 'Dickson on Poultry.' These were procured for me from Moscow, and answer the description well, except that the feathers on the legs are quilled, which they will probably lose in the next generation, our climate being so much milder than at Moscow."

According to Latham, this breed differs from others in having large tufts of brown feathers springing from each jaw, and some longer and fuller, like a Jew's beard, from the lower mandible. There is a tuft of upright feathers of the same silky texture,

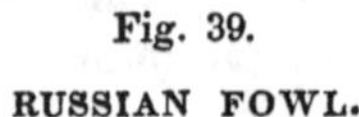

Fig. 39.

RUSSIAN FOWL.

springing from the backside of the head in the hen. The cock has a comb and wattles, the hen a comb only. This bird came from Moscow, and has fine variegated colors. The legs are covered with fine ordinary feathers. Some have the plumage of the game fowl, a fine tawny orange, spotted with black, and are highly esteemed in Scotland for prolific laying. This bird is called by some the Siberian Fowl. —*Dickson.*

Fig. 40.

OSTRICH FOWL.

This valuable variety, we have understood, first originated in Bucks County, Penn., hence they are called by some the "Bucks County breed." Some of this breed were first introduced into this vicinity some six or seven years since, from Philadelphia, by the late F. Bloodgood, Esq.

The specimens from which our portraits were taken, were presented to the author by a gentleman of Boston, who informed us he procured them from Maryland, where they were called the "Ostrich Fowl." In a letter accompanying the fowls he says, "This breed are the largest of fowls, and from them you will obtain the largest sized eggs. I have had eggs from this breed weighing 4½ ounces avoirdupois weight. I could have sold fifty pair if I had them to spare."

The color of the cock is a dark blue-black, with the ends of his feathers tipt with white; wings tinged with a bright yellow, or gold color; hackles dark glossy blue; rose or double comb, and wattles large; bold lively carriage and a stately walk. The hen does not differ much from the cock in color, and is very similar in form, being deep, short, plump, and thick-set in body; legs short, of a dark color and of medium size; she has a high, single, serrated comb, generally falling over on one side; wattles large.

This breed has one peculiar quality which we have discovered. When first feathered they are very dark colored; the white tips of the feathers are very small, and on moulting the white increases, and continues to increase with every successive moult until the white predominates. They are esteemed good layers, and for a large breed, good sitters and good mothers; the eggs large and nutritious; the flesh, unlike the Malay, white, firm, tender, and fine flavored. We consider them in all respects fully equal to the famous Dorking breed.

We are under particular obligations to Dr. R. Kitridge of Portsmouth, N. H., for the following information regarding the Booby Fowl, which, from his description, appears to be the Ostrich Fowl, under a different name. "Booby is a large fowl," says the

Doctor, "weighing from 6 to 9 pounds. Of those that I received, the smallest weighed 6 pounds, the largest 7½ pounds; the cock almost 9 pounds. Their invariable color is a black ground with white spots all over them; the legs are black; they are shaped like a turkey. They are great layers, and are not so much inclined to sit as the common hen; laying forty or fifty eggs before they are broody. I procured mine from Montgomery County, Pa."

In a letter from the gentleman in Philadelphia, who procured these fowls for the Doctor, he says, "I shall send two lots of fowls, a cock and three hens each. The Boobies are speckled, and were furnished by a German, and are no doubt a year old. There will be one hen with the three, that the good honest man said was much superior, and for which he was offered two dollars, on his way to the city. He had no name for this fowl, but said 'these are the greatest fowls ever in our part of the country.'"

The Portsmouth Journal gives an account of two varieties of hens, of more than three times the common size, and of proportionate value, which can be as easily raised as the common hen. "They have been raised by Dr. Kitridge, of that town." The editor says, "The kind called 'Boobies,' are speckled. The cock now weighs ten pounds, and some of the hens eight pounds. They are prolific layers; some of their eggs weigh over 3½ ounces each, and measure three inches in circumference."—*Am. Farmer.*

The editor of the Yankee Farmer says: "We have received from our friend Dr. Kitridge of Portsmouth, N. H., six Booby hen's eggs. These hens are considered as the greatest of layers by those who have kept them; and it appears that those which Dr. Kitridge has have laid well after getting over the effects of travelling. Owing to their being moved, they did not lay much for fifteen days; then they (four in number) laid thirty-six eggs in ten days, and none showed a disposition to sit excepting one, which he thought was not of the Booby breed."

TURKISH FOWL.

This is one of the most beautiful of domestic fowls, and is considered by Temminck and other writers, to be a variety of the Bantam, and distinguished by the variety and beauty of its plumage The body is whitish; the wing feathers and belly black; the tail black, tinged with iridescent green, and some of the feathers green on one side and black on the other; the body is streaked with golden and silvery streaks, adding much to its beauty. The legs and feet are blue. There is sometimes a fine crest of feathers behind the comb.

The Turkish hen is white, spotted here and there with black; the feet bluish; her wattles are smaller than those of the cock; in every other respect she is like the cock, except that her neck is yellowish, and tail all of one color.

Sir Wm. Jardine describes Indian fowls of the

Turkey breed as having a large comb, and tooth-cut on the front, wattles long and oval; the cheeks, throat, and fore-part of the neck, ending in a point bare of feathers; crown of the head around the comb, golden-yellowish brown,—the color also of the hackle feathers, which covers the whole back and sides of the neck; each hackle has a black centre; the fore-part of the neck is steel-blue around the naked skin. On the shoulders, the margins of the small feathers are pale golden-yellow with brown centres, edged on each side with black. The tail is nearly like our common fowls, but rather more horizontal, the centre feathers being longest, and curved and bending out.—*Dickson.*

GAME FOWL.

Buffon and other French writers term this the English Fowl; it is more slender both in body and in the neck, of genteel shape, giving the appearance of mildness; very keen bright eye, and the colors, particularly of the cock, are brilliant and showy. Their flesh is white, and superior in richness and flavor to all others; their eggs are small, fine shaped, and like their flesh are much esteemed for their delicacy. They are excellent layers and good setters. They lay very early in the season, and thus are prized by some as a useful cross with other varieties. A half-breed Game and Poland hen, it is said, will leave her chickens and begin to lay again in three weeks.

Buffon says a good hen will lay one hundred and fifty eggs a year (doubtful), and raise two broods of chickens besides. If any breed will do this it must be the above cross.

Fig. 41.

GAME FOWL.

Our portrait is copied from an English colored lithographic print, and called the "Shawl-necked" or "Irish Grey." They are the largest sized of the game fowls, and highly prized by the "fancy."

It is supposed by some that this breed originated from a cross between the common hen and the wild long-tailed Pheasant, as the latter is known to be of so quarrelsome and determined a character, that when two cocks encounter in their wild state, they seldom separate until one or the other is killed. The game breed certainly much resemble them in their plumage, color of their legs, &c., for the best are mostly red, dark brown, brass-colored wings, and black breast and body.

Mowbray says, "The progeny between the Pheasant and the common fowl, are necessarily *mules*, as proceeding from different species, although of the same genus."

Others have it that they originated, and which seems most probable, from the Bankiva Fowl, or Sonneretti's wild-fowl (Gallus Sonneretti). This bird, though smaller than the domestic breed, is superior in spirit and endurance, and usually proves victorious in combat.

That the Pheasant will cross with the domestic fowl is evident from the following fact, which we take from the Journal of Agriculture, published in Scotland. "In the autumn of 1826, a wanderer of the Pheasant tribe made his appearance in a small valley of the Grampians, the first of his family who had ventured so far north in that particular district. For some time he was only occasionally observed, and the actual presence of this *rara avis* was disputed by many; wintry wants, however, brought him more frequently into notice; and in due season, proof still more unequivocal became apparent. When the chicken broods came forth, and began to assume a shape and form, no small admiration was excited by certain stately long-tailed, game looking birds, standing forth amongst them, and continuing to grow in size and beauty, until all doubts of the stranger's interference with the rights of chanticleer effectually vanished. These hybrids partook largely of the pheasant character; and as they are of goodly size and hardy constitution, a useful and agreeable variety for our poultry-yards may be secured in a very sim ple and economical manner."

The Romans, whose taste for sanguinary specta cles is notorious, were extremely partial to the amusement of cock-fighting, and trained birds for the purpose. Indeed the taste for this cruel sport seems to be very general; the Mussulman nations of India are greatly addicted to it; the Chinese are devoted to the sport; and the natives of Sumatra enter into it with so much ardor, that instances, as it is said, have occurred of men staking not only their goods and money, but even their children, on the issue of a battle.

In England, as well as in this country, the same taste long prevailed, but happily the practice, "more honored in the breach than the observance," is now greatly on the decline; it is indeed incompatible with the diffusion of knowledge, the tendency of which is to humanize mankind, and lead the mind from sordid and debasing pursuits to sources of intellectual enjoyment.

"Philanthropists," says Mowbray, "are in the habit of declaiming much against the practice of cock-pit battles; but on reflection, the cruelty of that sport will be found among the least, wherein the feelings of animals are concerned, since fighting in the game-cock is a natural and irresistible passion, and can never take place against his will; and those that engage in regular combat upon the arena, would do so voluntarily, and with equal ardor, did they meet in the desert." But cock-fighting is regarded at the present day barbarous in the highest degree, and unworthy of the present enlightened age, and we are happy to say is among the things that have been.

Sportsmen who breed game-cocks for fighting are very particular in keeping them pure in blood, and have many curious names to designate them, such as piles, black-reds, silver-breasted ducks, brichin ducks, dark greys, mealy greys, black spangles, ginger duns, red duns, smoky duns, &c., &c., to the end of the chapter.

The game breed are *not* the fowls for the farmer or those who breed for the market; they are very quarrelsome, and their pugnacious disposition is manifested in the chickens at the earliest possible period. Whole broods scarcely feathered, are sometimes found moping about in corners, with sore heads and stone blind from fighting, to the very smallest individuals; the rival couples renewing their battles on obtaining the first ray of light. On this account few can be reared; and as this disposition, to a certain degree, prevails in the half-breed, it prevents crossing with the game-cock, otherwise a great improvement.

Fig. 42.

SILVER TOP-KNOT FOWL.

This variety is second only to the Golden Top-knot in brilliancy of plumage, and odd contrast of colors; being of a silvery-white ground with curious shaped black spots. In size they are between the Golden and the Black Poland. The feathers on the crown of the head are longer than the others, and their assemblage forms a tuft or bunch, the colors of which are mixed and the shape variable; those of the cock are rather an aigrette than a tuft, which gives them quite a lively appearance, and does not annoy their sight as it often does that of the Poland. The comb is double and very small, and their wattles are smaller than in other breeds. He has a bluish spot on each cheek, and a black collar under his throat, which gives him rather a dandy appearance.

The hen is considerably smaller than the cock, and is acknowledged, by all who have seen her, the most splendid bird of the gallinaceous tribe they had ever met with. Her colors are similar to the cock, about equally divided, and the dark spots have the appearance of scales. The crown or top-knot is unusually large, first rising from the head and then hanging over, which gives it more the appearance of a fine dahlia than anything else I can compare it to.

The specimens from which our portraits were taken (and I am sorry to say the artist has not done them justice), were presented to us by a friend residing on Staten Island, who said the breed was imported from France, and are said to be very prolific layers. They make a beautiful and showy appearance in the poultry-yard, and are very scarce in this country, nor are they described in any of the books on poultry.

In the neighborhood of Philadelphia one or two varieties of the Pheasant, such as the Gold and the Silver Pheasant, have been introduced, and are highly esteemed by those who have been enabled to rear them; but nothing is said about the top-knot. They are said to be a very beautiful bird, and reported prolific layers.

The breed of tufted fowls is that which the curious have mostly cultivated; and as it happens with all things that are closely looked into, they have remarked a great number of differences in them, especially in the color of their plumage, after which they have formed a multitude of divers breeds, which they esteem the more when the colors are more beautiful or more rare.

Fig. 43.

FRIZZLED FOWLS.

This fowl is said to be a native of Japan, and other parts of eastern Asia, and is frequently named the "Friezeland Fowl," from confounding the proper term *frizzled* with Friezeland. Capt. Steadman has observed, in his voyage to Surinam and the interior of Guiana, that the natives rear a very small species of fowls whose feathers are ruffled, and which seem to be natives of that country. It is distinguished by having all the ends of its feathers turned up and forwards, giving a singular ruffled or frizzled appearance to its exterior. It is frequently variegated with fine colors, but those which partake most of the original breed, have white plumage and smooth feet. The flesh is fine and delicate. It is less suitable than the common fowl for the farmer, the chickens being sensible to cold and wet, and are more frequently reared for curiosity than for other purposes. This bird has been introduced into this country, and by some we have heard it called the Russian and French breed, which is, however, a very different bird. Here, at the north, our climate is even too severe for the grown fowls.

The specimens from which our portraits were taken, were presented to us by our friend Dr. Wight of Boston, and are now in our possession, and we consider them of little value, further than making out our variety and gratifying curiosity.

Fig. 44.

SILKY FOWL.

This bird, by modern writers, is considered a species, rather than a variety. It is of good size, and the whole body is covered with feathers, the webs of which are disunited somewhat in the manner of some of the feathers of the ostrich and the peacock, and appear some like hairs and glossy silk. The legs are covered on the outside to the toes. Individuals of this sort differ in respect to color, as in other varieties, some are pure white, and others of a dingy-brown; and all of them with dark-colored legs, nor are the legs always feathered. This bird is indigenous in Japan, where it is much prized, and is also found in

China, where they are frequently offered in cages for sale to the Europeans. The skin and bones are said to be black, which gives it, when cooked, an unfavorable appearance, on which account it is in disrepute.

To propagate it in all its purity, requires that both the parents be covered with down.

This is the breed which gave rise, in 1776, to the fable of the rabbit-fowl, which was shown at Brussels, as the produce of a rabbit and a common hen, which was merely a downy fowl of Japan. It is said Buffon was for a long while teased by the letters of two pretended observers of Brussels, one of whom was a prebendary, and the other a Jew merchant; they were continually writing to him in order to convince him of the existence of the rabbit-fowl. Buffon had answered several times by arguments that proved the impossibility of such a disproportioned connection. Their credulous obstinacy at last put him out of temper, and he silenced them by a joke too bad to be inserted here, but which rid him for ever of the importunity of the Jew and the prebendary.

We have seen fowls of this description, and once bred one from common fowls. Our portrait was copied from Parley's Universal Geography.

Fig. 45.

BANTAM FOWL.

This is a small but beautiful variety, prized for its singular and grotesque appearance, but more for its prolificness in eggs and for the delicacy of its flesh. A full bred cock of this breed should not weigh more than one pound. They make a good substitute on the table for a partridge. The cock should have a rose comb, wattles small, beak and feet yellow, and a proud lively carriage. They are elegantly formed, and of all colors. The nankeen-colored bird, his feathers edged with black, his wings barred with purple, his tail feathers black, his hackles slightly studded with purple, and his breast black with white edges to his feathers, and those all black, or all white, without the admixture of any other color, is most valued. The hens should be small, clean legged, and match in plumage with the cock. Some of this breed are feathered to the toes, so as to obstruct their walking; though feather-legs are not exclusively peculiar to Bantams; for Bantam-fanciers, with Sir John Seabright at their head, preferred those which had clean bright legs, without any vestige of feathers.

This variety has received its name from the district of Bantam in Java, from whence it was brought by the Dutch. The eggs are small, like those of the partridge, but of the best quality. The Bantam must, however, be considered more as an object of curiosity than utility, and of course expect to be received with no peculiar favor in this country.

Aldrovand, two thousand years ago, describes the cock with the neck and the back of a chestnut color, the wings at first black, with whitish spots, afterwards black, the quill-feathers being white on the outer and black on the inner sides; the throat, breast, belly, thighs and legs, black with whitish spots; the feet yellow; the wattles large; comb double, and not very large; the beak yellow; the tail-feathers partly white and partly black. The hen is of a yellowish color, and everywhere, except the neck, marked with oblong black spots.

Mowbray says, "There has lately been obtained a variety of Bantam, extremely small and as smooth legged as a game-fowl. From their size and delicacy, they are very convenient, as they may always stand in the place of chickens, where small ones are not otherwise to be had. They are also particularly useful for sitting upon the eggs of pheasants, being good nurses as well as good layers."

There is also a South American variety, either from Brazil or Buenos Ayres, which will roost in trees. They are said to be very beautiful, partridge spotted and streaked; the eggs small and colored like those of the pheasant; both the flesh and eggs are fine flavored and delicate.

The white Bantams, with long feathers on their legs, are the most common, but our portraits were taken from some rare fowls obtained of a neighbor, which are highly prized for their great beauty.

"One of the prettiest little Bantam patriarchs we have seen," says Boswell, "was on a visit to one of the finest landscape painters of the day, in the yard of our friend Mr. Brown. He marched majestically at the head of his tiny tribe, and was of a very fine breed from Ayrshire. They had the full scope of the garden, and did little injury—the door-step was their feeding place, and still did no discredit to the tidiness of good old Bernie, so that two or three Bantams may be kept without much molestation in any rural situation."

In the American Agriculturist, a writer under the signature of L. F. A. says, "The Bantam is a beautiful little bird, usually white in color, with short legs, feathered oftentimes to the extremity of its toes. It is often of variegated colors, inclined to red, brown and white, prettily mixed. Occasionally a variety is met with that are smooth legged. They are very domestic, often making their nests in the kitchen and cupboards of the dwelling when permitted. They are excellent layers, and good nurses; but require a dry location, on account of their short, feathered legs

The males are wonderful crowers, and exceedingly pugnacious. They arrive at maturity early, and are well worthy of propagation."

The Bantam cock is very courageous, and will even make battle with a turkey. The feather-legged kind may often be kept, for fancy and amusement, especially where there is convenience for no other kind, as they are not so apt to scratch and do injury to the garden.

The Bantams are the fowls of all others for the city. We have known them to prosper and lay well through the winter in a cellar well lighted. They are very tame and domestic, faithful sitters, good mothers, and will lay more eggs, though small, than any other variety. They require but little food, and thrive cooped up in a small yard where there is dry sand or ashes and sun.

The following remarkable instance of the attachment of a Bantam cock to his mate, we find related in a late English publication. Speaking of the cock the author says—"He is also capable of such attachment to his mate, that we remember a Bantam cock and hen which were kept for some years as favorites, without any others, in the stable yard of our father, and when, at length, the hen died, the cock, seeing her lifeless, but naturally unconscious of its being a final separation, hovered around her, calling to her, and pecking at her gently, as if to awake her. Though corn was offered to him, he refused to eat, or to roost at night, but moped round the yard, vainly searching for his old companion, when not finding her, he flew away, and was never after heard of."

Fig. 46.

JAVA BANTAM.

Since writing the foregoing, we find in a late London paper the following account of her majesty's Java Bantams: "These birds are perfectly white, but present no peculiar characteristics of form or structure; their habits, however, are in some respects so singular as to demand especial notice. The cocks are so fond of the hen's eggs, that they constantly break and suck them; and they have been known to attack the hen, to tear open the ovarium, and eat the egg-shells. To subdue this propensity, her majesty's keeper gave the cocks first a hard-boiled egg, and then a marble one to fight with, taking care at the same time to keep them from any access to a real egg. No sooner was this done, than an attack on the false egg was commenced, which lasted for a week, till at last, wearied with their fruitless labor, they gradually gave up all notice of them, and with that abandonment as was anticipated, they ceased from their accustomed destruction of the eggs, and have never been known to attack them since."

They are represented as wonderful layers; and in proper season their fondness for offspring is so strong, that on a trial of its capacity, a hen in her majesty's possession, sat a period of nine weeks on three successive sets of eggs.

Among her majesty's fancy fowls, it is said, are to be found some splendid specimens of Sir John Seabright's breed of Bantams, a cock of which, remarkable for his martial bearing, is a great favorite with Prince Albert; also a fine collection of Scotch Bantams, including some curious "crosses" with grouse-birds; and several frizzle fowl, remarkable for their white, silky hairlike feathers and their black skins.

BUCKS COUNTY FOWL.

There are so many kinds of fowls called Bucks County breed, that we find it quite difficult to know, much more to describe the real "Simon pure." In some sections the Ostrich (No. 40) are called Bucks County breed, and in others a large, coarse, dark-colored, swaggering fowl, of which our portraits are a fair representation, and in our opinion (which, however, is founded only on one year's experience), have little to recommend them except great size. They are enormous eaters, poor layers and miserable sitters.

In corroboration of our opinion hear the testimony of others. Thomas P. Hunt, in the New England Farmer, says, "The large Bucks County hens, weighing as much as the Malays, are not good

Fig. 47.

BUCKS COUNTY FOWL.

layers, and their eggs are very apt to have double yolks; of course do not hatch well."

Thos. P. Thurlow, of Pennsylvania, who has paid some attention to poultry, in a letter to the author, says, "As far as I am acquainted with the Bucks County breed, they would not do to lay, as they very seldom lay more than ten or twelve eggs at a litter. They are profitable to breed from, provided you can make capons of them. They are often sold in the Philadelphia market from four to five dollars per pair, but they eat a great deal more than the common fowls."

Now hear what our friend L. F. A. says of this breed in the American Agriculturist. "The Bucks County breed has received some celebrity in the neighborhood of Philadelphia, as a valuable variety of fowl, principally on account of its enormous size. I have seen many specimens of this fowl, paid some attention to its habits, and learned from those who have tried them, their principal merits. It is a large bird, weighing at maturity eight, and even ten pounds; rather thinly feathered, of various colors, from grey to black, and frequently speckled black and white. They are coarse in their legs, tall and bony, and have evidently a cross of the Malay in their composition. They are but moderate layers; their eggs large and good. They are bad sitters, frequently breaking their eggs on account of their great weight and size, by crushing them; are not hardy, and, on the whole, will not compare with the common dunghill fowl for ordinary uses. They do not breed *equally* in size and appearance, showing them evidently to be a cross from other breeds; but from what they are derived, other than the Malay, it is difficult to say. A gentleman of my acquaintance, who is very curious as well as nice in the selection of his fowls, tried them effectually for his poultry-yard, and they disappointed him. He then crossed them with the game breed, and has succeeded finally—the cross being reduced in size, fuller feathered, hardier, and better layers, with an excellent carcass and finer flesh. As a fancy fowl, or to make up a variety, they are very well; but can never become of great *utility*, except to cross with the common or the game fowl, to the farmer."

Fig. 48.

RUMPKIN FOWL.

Some writers, among whom is Temminck, consider this bird a distinct species rather than a variety; the wild breed from which it originated still existing, and confined to the deep forests of Ceylon. Its principal characteristic is the want of a tail, hence its name, Rumpless Fowl, by which it is more commonly known; though in the wild bird the comb is not indented, and the wattles are blood-colored rather than scarlet. This bird was early domesticated in this country, and from the fact of their being early seen in Virginia by travellers, Buffon imagined they were indigenous there. One of the most singular facts about this bird is its entire destitution of the gland on the rump, from which it is supposed common fowls derive their stores of oil to smoothe their feathers and protect them from rain. As the feathers of the Rumpkin are not less smooth than those of other fowls, and shed rain equally well, it would seem this gland was destined to perform some other office, and that the common notion on this subject had little foundation in fact. In the wild breeds the feathers are all of a dusky orange, and this, on a reddish brown, seems to prevail among the domesticated ones. There are, however, some instances in which the common birds are finely variegated—one in the possession of the author, from which our portrait was taken.

This breed is looked upon by some to be a native of Persia. Buffon thinks, on the contrary, that Virginia is the place whence it sprang. He grounded his opinion on the one hand on what is reported by the Philosophical Transactions of 1693, that when fowls are led to that country they seem to lose their rumps; and on the other, on naturalists having only begun to mention fowls without tails after the discovery of America. I am not of that opinion, says Main, which appears not admissible. In fact, modern travellers have not confirmed the loss of the rump which the English experience in Virginia, and it is positively known, that in the other parts of America, in the hottest even, this privation does not take place.—*Main.*

Aldrovand describes the cock as black, interspersed with yellow streaks, and the chief wing feathers white, the breast white, the feet ash grey. The hen has a smaller comb than the cock, and is of a rusty color except three black feathers in each wing.

Fig. 49.

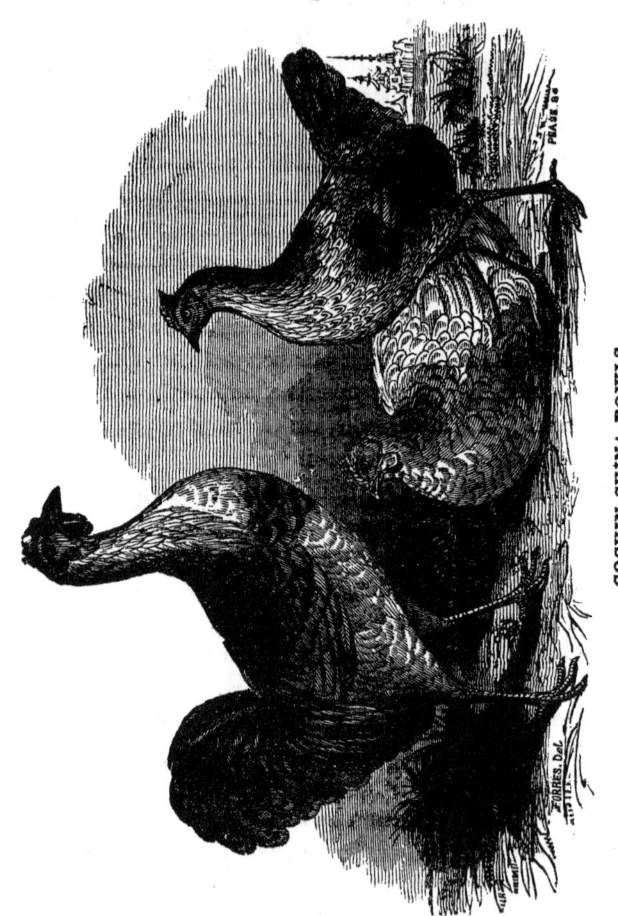

COCHIN-CHINA FOWLS.

In the "London Illustrated News," we find a description of Queen Victoria's poultry-houses, yards, and portraits of some of her rare fowls, among which are some giant birds, termed "Cochin-China Fowls," which we have transferred to our pages. From the portraits and description of them we are reminded of some of the same breed which we saw in the yard of Mr. Geo. Law, in Baltimore. They were noble looking birds, and as large as turkeys. The cocks were of a red color, and the hens of a yellowish brown. We saw nothing, however, to recommend them except their great size.

"Her majesty's collection of fowls is very considerable, occupying half-a-dozen very extensive yards, several small fields, and numerous feeding-houses, laying-sheds, hospitals, winter-courts, &c. It is, however, in the new fowl-house that the more rare and curious birds are kept, and to those—as the common sorts are well known,—we shall confine our attention. The Cochin-China fowls claim the first consideration. These extraordinary birds are of gigantic size, and in their proportions very nearly allied to the family of bustards, to which, in all probability, they are proximately related—in fact, they have already acquired the name of 'Ostrich Fowl.' In general color they are of a rich glossy brown; tail black, and in the breast a horse-shoe marking black; the comb double. Two characters appear to be peculiar to them—one, the arrangement of the feathers on the back of the cock's neck, which are *turned upwards;* and the other, the form of the wing, which is jointed, to fold together, so that, on occasion, the bird may double up its posterior half and bring it forward between the anterior half and body. The eggs are of a deep mahogany-color, and of a delicious flavor. These birds are very healthy, quiet, attached to home, and in every respect suited to the English climate. They are fed, like most of the other fowls, on a mixture of boiled rice, potatoes, and milk."

In addition to the foregoing varieties, we find in Main's "Treatise on Poultry," and in "Buffon's Natural History," the following, which are probably sub-varieties, produced by crossing and inter-crossing, having some peculiarities of tuft, color, &c., which marks the bird, while the general qualities are the same.

The Adrian Fowl.—The ancients so named a dwarf breed, which they got from the environs of Adria, a town in Italy, which had given its name to the Adriatic sea. Aristotle speaks in high terms of the fecundity of these fowls; "they lay," says he, "every day, and sometimes two eggs a day."

The Sausevere Fowl.—Tavinier has seen this fowl in Persia; it is described as being a very large breed, the eggs of which sell for three or four crowns apiece, and which the Persians amuse themselves by breaking one against another, as our boys do "paas-eggs." A fine cock of this breed sells in Persia, according to the same traveller, as high as three hundred livres.

The Alexandria Fowl.—The ancients mentioned this as one of the finest breeds; it has now nothing remarkable in it.

The Silvery Fowl.—This is the name which the curious have given to tufted fowls, the plumage of which exhibits spots regularly distributed, and of a very clear white.

The Carux Fowl.—This is almost twice as large and bulky as the common fowl, from which it does not differ otherwise. The chickens of this breed get their feathers later than those of the common breed.

The Lombardy Fowl.—A very small breed of fowls, which some travellers have spoken of as being natives of the Isle of Madagascar, where it is called "Acoho."

The Media Fowl.—Which commentators have improperly termed "Melos Fowl," by reading gallicus melicus, for *gallus medicus.* A large sized race, and the males accounted courageous by the ancients, but the females of which are not productive.

The Philippine Fowl.—Independent of the Cambogia fowl, which the Spaniards have taken to the Philippines, there exists in these islands another breed, that goes by the name of Xolo, and which has very long legs.

The Rhodes Fowl.—A large breed very much esteemed by the ancients. The cocks stronger than the others, were kept for fighting; but these birds, which had so much ardor to fight, had very little for their mates; three hens only (instead of fifteen or twenty), were enough for one cock; and not such good sitters as the common hens.

The Barbary Fowl.—This African variety is generally of a pale or dun color, spotted about the neck sparingly with black, and the feathers at that part very full; on the crown is a large tuft of feathers, the same in color with the body.

The Marcellus Fowl.—The Marcellus or French hen is a good layer, medium size, very large comb, white legs, plumage various, but often white, with dark heads, tail and wings.

The Five-clawed Fowl.—The character of this breed is having five claws to each foot, three before and two behind, like the Dorking fowl.

The Half-India Fowl.—The Dutch have given this name to a breed of fowls peculiar to the Island of Java, the males of which carry their tail something like a turkey. These cocks are seldom bred except for fighting. This is probably the Java fowl.

The English Dwarf Fowl.—A very small fowl which has been greatly multiplied in England, because it is very fruitful, and excellent for sitting; it is preferred in pheasants' walks to the common hens, which are too heavy. When the breed is pure, the plumage of this fowl is quite white; it is not larger than a middling sized pigeon.

The Chinese Dwarf Fowl.—Smaller than the English Dwarf Fowl; its plumage is varied on the different individuals like that of the common breed.

The painting of them is frequently to be found on China papers.

The French Dwarf Fowl.—A small breed of fowls, not so small, however, as the English Dwarf Fowl. Its plumage varies like that of the common breed; its legs are very short, and its eggs are not larger than that of the pigeon.

The English large-footed Fowl.—A variety of large-footed fowls, distinct from the Bantam fowl and which is larger than the French one.

The French large-footed Fowl.—Its legs are covered with feathers as far as the claws. The greater part of the large-footed breeds have no tuft.

The Fowl all black.—Besides the Negro Fowl, there exists in some parts of Africa, and at Sumatra, another breed still blacker, since its very bones are black as jet. Marsden makes the distinction of this fowl with the Negro Fowl, which is equally found in the island of which he has written the history.

The Widow Hen.—Small white dots strewed on a dark ground, have made this name be given to a variety of tufted fowls.

The Hamburgh Fowl.—The Hamburgh cock, named also velvet-breeches, because its thighs and belly are of a soft black. Its demeanor is grave and stately, its iris is yellow, and its eyes are encircled with a ring of brown feathers, from which rises a black tuft that covers the ears. There are other feathers nearly like these behind the comb, and beneath the bar-bills, and broad round black spots on the breast. His legs and feet are of a lead color, excepting the soles of the feet, which are yellowish.

Ever-laying Fowl.—A writer in the Quarterly Journal of Agriculture, says, " Any of the breeds of domestic fowls may become what is termed ever-laying, that is, not inclined to hatch, an artificial temperature produced by domestication."

We are indebted to Mr. Sanford Howard, a practical agriculturist, and now one of the associate editors of the Cultivator, for the following history and description of a valuable breed of fowls once owned by him:

" Mr. Bement—The first I ever saw of the variety of white fowls, about which I spoke to you, was at the country seat of the Hon. John Wells, near Boston, some twenty years ago. They much attracted my attention as I frequently passed the place; their large size, fine shape, and clear white plumage, showing them off with beautiful effect as they walked over the green lawn. I finally spoke to Mr. Wells about them. He told me they had been brought by one of his vessels 'from over the seas,' but from what place he could not tell. He said their flesh was of excellent quality, but he did not think them as good for laying as some others.

" Mr. Wells afterwards gave me some of these fowls, which I took to Maine. I soon found that their quality was all that Mr. W. had said of them; and after the breed had become a little known, they would readily sell to those fond of good eating, for about twice as much per pound as common chickens. They are small boned, rather short in the legs, and have an uncommon tendency to fatten. They will weigh, when pretty well fed, from four to four and a half pounds each when dressed, at five months old. Their flesh is very white, fine grained, and high flavored; and unlike most fowls, their flesh does not grow tough and hard with age. I have seen them sufficiently tender for the spit at the age of two years, though I cannot say they would *always* be so. They are quite middling as layers, until they are past two years old, when they usually become so fat that they lay less.

"When I went from Maine to Ohio in 1837, I let Payne Wingate (a member of the Society of Friends, and a man well known for his nice attention to many departments of rural economy) have my fowls. When I was in Maine in '42, I found friend Wingate in possession of a fine stock of them. I think he had rather improved them, after they had come into his possession. He told me he had killed some at five months old which weighed five pounds each, dressed, and that they were as much esteemed as ever for the table.

"In 1842 I took some of these fowls to Ohio. H. S. Sprague of Newark, and Isaac Dillon of Zanesville, in that State, now have them.

"Yours truly, SANDFORD HOWARD."

CHAPTER IX.

Pea-fowl—Wild Turkey—Domestic Turkey—pairing—laying—hatching—food—fattening—Guinea Hen.

Fig. 50.

THE PEA-FOWL.

THIS most magnificent and beautiful of all the feathered race, is supposed to have been originally a native of India; but they have been long introduced into Europe and this country, as ornaments to the mansions of gentlemen farmers. Peacocks are said to be at present found in a state of freedom upon the

Islands of Java and Ceylon. The history of King Solomon is an evidence of the antiquity of the Peacock, and also the choice of the goddess Juno, who selected this for her favorite bird, from its gorgeous and brilliant plumage and majesty of demeanor. It is asserted by the ancient writers that the first Peacock was honored with a public exhibition at Athens; that many people travelled thither from Macedonia to be spectators of that beautiful phenomenon, the paragon of the feathered race. Exclusive of the consideration of ornament to the poultry-yard, the Peacock is very useful for the destruction of all kinds of reptiles; but at the same time some Peacocks are said to be vicious, and apt to tear to pieces and devour young chicks and ducklings suffered to come within their reach, on account of which we have discarded them from our premises.

When young they are a tender bird, and show all that constitutional delicacy which bespeaks their tropical origin; but after they are six or eight months old, they become inured to our northern climate, and live and roost among the common poultry. In the summer months they rather choose to roost on trees or on the top of buildings; always making use of their wings more than other poultry when inclined for flight.

There is a white variety of this bird, but they are rarely met with. We once saw one at the residence of C. H. Hall, Esq., at Harlem.

It is said they live to the age of twenty years, and

at three, the tail of the cock is complete and gives its full splendor. Their color is too well known to need description, and if not, my pen could not do it justice.

They are not worth the attention of the farmer on the score of profit, but they may be useful to keep watch; as they will roost on the highest chimney, the top of the barn, or any elevated place, and from it they will issue their piercing cry on the approach of any stranger or enemy, taking the place of the watch-dog.

The crow, or rather scream, of the Peacock is loud, harsh and disagreeable. It is only heard during the breeding season; and it is one of those rural sounds which proclaim the approach of summer. This, together with the frequent appearance of the cock in "full glory," exhibiting his splendid train, are sure signs that the business of nidification is at hand. The hen has always much apparent listlessness in her manner; for even when looking about for a proper place to nestle in, she makes no sign that she is so engaged, but walks leisurely about, as if she was looking for food. She is, however, seeking the most private corner she can find, at some distance from the concourse of houses of the other poultry. If a wood or shrubbery be near, she will choose a place under a thick bush, and generally among dry fallen leaves. As they cannot easily be persuaded to adopt a nest provided for them, like some other poultry, the hens require to be strictly watched during the laying season, to find the nest, in order that the eggs be removed as soon as laid, substituting for them pieces of chalk cut in the form of eggs. The hen always covers her eggs on leaving her nest, and the same device must be practised in laying in the chalk eggs. If the place she has chosen be secure from foxes and other wild animals, hogs, strange dogs, or other annoyance, the hen may be allowed to sit there, because they generally hatch well when allowed to have their own way; but if the place be unsafe, she should be set in the house, in a nest made for her, the eggs put in and shown her, and which, sometimes, she will readily take to; but if she be obstinate, and will not sit, the eggs must be put under a common hen, which will perform the part of an excellent step-mother.

We have known, however, a fine brood of pea-fowls hatched by the mother in an awkward place, at a considerable distance from the homestead, and at last brought home by the cunning mother to join the other poultry in the yard. The hen was observed to come every morning to the yard, and with importunate clamor demand her breakfast, after which she would take wing and fly away out of sight.

The best food for young pea-fowls, are ant-eggs, as they are called, barley-meal paste mixed with sweet curd and hard boiled eggs chopped fine; when grown up they live on any kind of grain. The ant-eggs are found in woods, and taken from the nests of the large house-ant (*formica herculeana*). These insects form considerable hillocks of the hybernaceous leaflets of the beech and other trees, and the smallest bits of withered twigs of the same. In the centre of the hillock the eggs are stored, which being dug out and thrown upon a cloth, are tied up in it and carried to the poultry-yard. A handful or two of this, the natural food of the birds, twice a day, will be sufficient, along with an occasional feed of the other ingredients above specified.

The chicks should be housed in wet weather; and every day, when dry and warm, should be allowed to divert themselves on smooth turf. If hatched by their own mother, she must be cooped for a fortnight or three weeks, to prevent her from rambling too far from home, which they are apt to do if not confined.

This bird has long ceased to form a common dish for the table; and probably, from the coarseness and dark color of its flesh, when it did, the motive was rather show than use; but pea-hens and pea-chicks still retain their place at some of the great feasts in particular parts of Europe.

Pea-fowls being chiefly valued for their beauty always bear a high price alive, in market. As they are always allowed the use of their wings for their own protection, they are often sad pests in the garden, if once they find their way thither. They will devour a large crop of strawberries as fast as they ripen, if not kept off.

Fig. 51.

THE WILD TURKEY.

Next to the common fowl, Turkeys form the most numerous tribe, and for the table one of the most important and useful birds of the farm-yard. It is a native of this country; and we can boast of the Wild Turkey, a bird so truly valuable, that Dr. Franklin observes, "It would have been a much fitter emblem of the country than the white-headed Eagle, a lazy, cowardly, tyrannical bird, living on the labors of others, and more suited to represent an imperial, despotic government, than the republic of America." However true this may be, the Turkey is entitled to the nobility of the barn-yard.

The Turkey was unknown before the discoveries of Fernandez. They were first introduced from Spain into England as early as 1525, and were in a short time spread over the whole kingdom, and increased to that degree, that in 1585 they could already furnish a dish in country feasts. The Turkey did not reach France quite so early; the first eaten appeared at the nuptial feast of Charles IX., in 1570. They have since been domesticated throughout the civilized world, in every climate, although said not to succeed equally on the barren sands of Africa.—*Boswell.*

Buffon remarks, "As the Turkey was unknown before the discovery of America, it has no name in the ancient language. The Spaniards called it *pavon de las Indias*—the Peacock of the Indies, because its tail is like a Peacock."

They were also called the Indian Cock and Hen, as they were first taken from the West Indies to Europe. "It is difficult," says Boswell, "to ascertain the etymology of their present name. It appears to have been introduced as a satire upon the solemn strut of the bird, which might appear to give it a resemblance to the pompous stride of a Turk.

Most assuredly it did not arise from the native place of the bird, which has no connection with eastern Europe or Asia. To suppose the bird Melagus, mentioned by the Greek writers, to have been the Turkey, is quite a mistake. When first discovered in America, they were seen both in a wild and domestic state."

"The great size and beauty," says Audubon, "of the Wild Turkey, its value as a delicate and highly prized article of food, and the circumstance of its being the origin of the domestic race now generally dispersed on both continents, render it one of the most interesting of the birds indigenous to the United States of America."

Audubon says, "the flesh of the Wild Turkey is of excellent flavor, being more delicate and juicy than that of the domestic Turkey. The Indians," continues he, "value it so highly that they term it, 'the white man's dish.'" Although they are declared to be of excellent quality, still there is a tradition, as old as the time of the Buccaneers, that they were not worth powder and shot.

The plumage of the Wild Turkey is generally described as being compact, glossy, with metallic reflections; feathers double, as in other gallinaceous birds, generally oblong and truncated; tips of the feathers almost conceal the bronze color. The large quill coverts are of the same color as the back, but more bronzed with purple reflections. The lower part of the back and tail coverts are deep chestnut, branded with green and black; the tail feathers are of the same color, undulatingly barred and minutely sprinkled with black, and having a broad blackish bar towards the tip, which is pale brown and minutely mottled; the under parts duller; breast of the same color as the back, the terminating black band not so broad; sides dark colored; abdomen and thighs brownish-grey; under tail coverts blackish, glossed with brown, and at the tips bright reddish-brown.

The plumage of the male is very brilliant; that of the female is not so beautiful. When strutting about with tail spread, displaying himself, this bird has a very stately and handsome appearance, and seems to be quite sensible of the admiration he excites.

Dr. Bachman says, "that in a state of domestication, the wild Turkeys, though kept separate from tame individuals, lose the brilliancy of their plumage in the third generation, becoming plain brown, and having here and there white feathers intermixed."

The unsettled parts of the States of Michigan, Ohio, Kentucky, Illinois, and Indiana, and an immense extent of country to the north-west of these districts upon the Mississippi and Missouri, and the vast regions drained by these rivers, from their confluence, to Louisiana, including the wooded parts of Arkansas, Tennessee, and Alabama, are the most abundantly supplied with this magnificent bird. It is less plentiful in Georgia and the Carolinas, and becomes still scarcer in Virginia and Pennsylvania, and is now very rarely seen to the eastward of the last mentioned

States. In the course of my rambles through Long Island, the State of New York and the country around the Lakes, I did not meet with a single individual, although I was informed that some existed in those parts.

I have since ascertained that some of these valuable birds are still found in the States of New York, Massachusetts, Vermont, and Maine. In the winter of 1832-3, I purchased a fine male in the city of Boston. This species is abundant in the wooded portions of Texas, but none have been observed either on the Rocky mountains or to the westward of them. They are, however, becoming less numerous in every portion of the United States, even in those parts where they were very abundant thirty years ago.—*Audubon.*

We have been informed by Prof. Emmons, that he understood a flock of fifteen or sixteen wild turkeys had been frequently seen in the town of Chester, Mass., and that he saw one that was shot there in 1841.

A lot of these live wild turkeys were, in the winter of 1843, brought to the Albany market, from the west, and sold to one of the inn-keepers, one of which is now in the possession of a gardener, who is crossing him with some of our tame varieties, to improve the size and hardihood of the domestic breed.

The wild turkeys are described as being much larger than the tame ones. Far from being improved by care and abundance of food, contrary to most other wild animals, this species have degenerated.

Wild turkeys, it is said, often weigh from forty to sixty pounds.

In a state of nature, they are said to parade in flocks of many hundreds, feeding, in general, where nettles are to be found, the seeds of which are their common food; they also feed on acorns, beech-mast nuts, &c., and become so fat as to be unable to fly more than a few hundred yards, and are then soon run down by dogs and horsemen. They roost in the highest trees, and, like our partridge, a number on the same tree can be easily shot, by taking the lowermost ones first. Since the settling and cultivating such extensive tracts, the wild turkeys, like the deer, the buffalo, and the Indians, have been driven into the uncultivated regions, and have become very rare, and, at no distant period, will in all probability become extinct, as the cock-of-the-woods, as well as the bustard in Britain.

Where there is an abundant crop of acorns, there immense flocks of turkeys may be expected. In the autumn, they direct their course in vast numbers to the rich lands on the borders of the Ohio and Mississippi. The males and females travel separately, but all in the same direction. Before crossing a river, they assemble on the highest eminences, and remain there as if in consultation for a day or two. At length, after due preparation, the leader gives a signal note, and they all wing their way to the opposite shore. Some of the young and weak ones fall into the water, and are obliged to swim for their lives, using all their means and power, and the most violent exertion, to reach the shore. Many, however, perish in the attempt. It is observed after these journeys, the turkeys are so familiar, that they fearlessly enter the plantations in search of food. Great numbers are killed at this time, and kept in a frozen state to be sent to distant markets.

The author, on a visit to the farm of Col. Jaques, near Charlestown, Mass., in the winter of 1842, saw a flock of some ten to fifteen of this splendid bird, which were apparently as tame as the common turkey. They were very large, of beautiful plumage, and the cocks were strutting about with as much pride as a nabob.

The following very interesting letter was published in the third No. of the 10th Vol. of the Cultivator, and we cannot resist the temptation to insert it here. It is dated,

" *Rockville, Ill., January,* 1843.

" Ever since General La Fayette was in this country, and expressed a wish to obtain some wild turkeys to take with him to France, I have felt an interest in this kind of poultry. When I came to this State in 1835, I made up my mind that I would endeavor to obtain and domesticate some of them as soon as I could. Upon inquiry, I learned that occasionally a nest of eggs was found and put under hens to be hatched out, and the young ones killed in the following winter, under an apprehension that they would run off in the spring. In 1837, a nest of eight eggs was found, which I purchased. These I set under a hen, and all were hatched out. A difficulty now occurred. The young turkeys paid no attention to the hen's cluck—were disposed to wander about, and make the hen follow them, instead of their following the hen. In this way they were in danger of being lost, and some of them were lost in consequence of going further from the house than the hen was disposed to follow. Of the eight, I succeeded in raising but one, though I believe some of them died a natural death. This one kept company with the hens and roosted with them, till by repeated alarms from vermin, he forsook the hen-house, and took the roof of a stable for a roosting-place. This proved to be a cock, but it was not till he was more than eighteen months old that he began to strut and gobble. In a very windy time, he would leave his roosting-place, and go into the hazel bushes which were near by, to lodge. On one of these occasions he was killed, probably by a prairie wolf, as they were plenty in the neighborhood. This turkey was so thoroughly domesticated that he would eat corn out of my hand, and showed no more disposition to wander off than my hens.

" The last year a wild turkey's nest was found in the neighborhood, and I obtained seven of the eggs, and set them under a hen; six of them hatched, and I made a yard on a grass-plat for them, by setting up boards edgeways, in a square form, and one tier high, where I kept them confined till they became attached to the hen, and then let them run at large. I took some pains to have them run in the garden to keep the vines clear from bugs, which they did effectually, without doing any damage to the vegetables. My place, as I keep no dogs for the protection of poultry, is beset with what our folks call the *wild varmints*, such as weasels, skunks, minxes, raccoons, and prairie wolves, so that I succeeded in raising but three out of six. These I valued very highly, and thought them out of danger, when a hunter came along, and seeing them a little distance from the house, shot two of them, so that my stock is reduced to one solitary hen."

My own experiments in rearing wild turkeys go no further, but I know of the results of the experiments of others of my acquaintance in this county, some of which I will give. A family in a small grove, had a pair of wild turkeys; the hen, when a year old, laid a nest of eggs, but none of them hatched; a season after this, she raised a litter of young ones, which, after getting their growth, with the old ones in company, left the grove, and never returned. Another family had a pair; the hen got accidentally killed when about a year old, and the male kept company and roosted with the fowls. The next spring, the owner procured two tame hen turkeys, and succeeded in raising twenty-eight half-bloods. A neighbor had a wild male domesticated, one year old, and two tame hen turkeys. That season not one egg hatched; the next season, from one of the hens, the other having died, he raised fourteen or fifteen half-bloods The cocks (unlike the tame in that respect), it seems, do not come to maturity till they are two years old. The half-blooded turkeys bear a greater resemblance to the wild than they do to the tame, and will breed equally well.

Although my attempts at rearing wild turkeys have been so unsuccessful, I am not discouraged, but intend to persevere in the effort a while longer; I believe it will be some little object to raise the full and half-bloods here, but much more of an object in your neighborhood, for the city market. The reason why they would have a preference over the tame turkey, is, 1st. That they are larger; one was killed about ten miles from me which weighed twenty-four pounds; and I have heard of their being killed in Ohio, one at least, which weighed rising of thirty pounds. 2d. They are more robust, will bear the rain and wet grass, and therefore more easily raised than the common turkey. 3d. They are hunters of flies, bugs, and other insects; and eat much less corn or meal than other poultry, and are of course more profitable.

"They are a more beautiful fowl than the tame Turkey, their color being brown or snuff, with feathers having a lustre or brilliancy upon them in some degree like the peacock. There are at present, within five miles of me, five full blood wild turkeys. If the males were in your part of the country, I should think them better worth $50 a piece, to run with a flock of tame hens, than some imported animals are worth the prices they have brought."

Fig. 52.

DOMESTIC TURKEY.

We have spoken of the turkey of nature; we will now treat of the turkey of art; that is the turkey that makes so interesting a part of our rural and domestic economy. They are, next to the common fowl, the most useful and valuable of domestic birds, and at the same time that which requires the greatest care in the first moments of its existence; when once reared, however, every temperature agrees with it.

Of the turkey, Buffon and others assert that there is but one species; and in this country we have three varieties, the black, the copper-color, and white. The first is considered the most hardy and generally preferred; the second is held by some in great esteem; the latter, with their red head and caruncles, contrasted with their snowy whiteness, make a pretty appearance and add much to embellish the grounds of the house, and their feathers are valuable, and they have much down on the thighs.

According to Parmentier, the white turkey, contrary to analogy, is by some thought to be more robust and easily reared and fattened; and hence large flocks of these may be seen in some parts of France. Such, however, is not our experience, but the reverse. The black turkey, on the contrary, is always most marketable, from its being said their skin is white, and their flesh finer and sweeter; while the males are larger, and the females are better breeders.

There can be little doubt that black turkeys are produced in greater numbers than any other color. Madame Clavier, an ingenious French lady, fond of rural economy, told M. Parmentier that she had a white turkey cock in her yard, with ten black turkey hens, and yet she never had a white chick hatched, nor even shaded in the slightest degree. In Dauphiny, on the other hand, they are found of every shade of color, from a deep black to a pure white.

Mowbray tells us, that a turkey cock, the property of a gentleman in England, which was black in 1821, became afterwards perfectly white, and in the process of moulting, just before this singular change, it gradually showed every shade between the two colors, the feathers being alternately black and white.

"Although not of ancient date," says Main, "the subjugation of turkeys has already produced varieties in our (England) climate. The most remarkable is that of the *tufted turkey*, as yet very rare, and whose tuft is sometimes black and sometimes white."

Turkeys, in the neighborhood of large woods, if not watched and prevented, will eagerly stroll thither without any design to return, such is the natural wildness of their species. In corroboration of the above, an instance was communicated by a correspondent in the Sporting Magazine, of May, 1824, who says, "Two years last harvest, a two year old cock, and two hens of the same, belonging to me, were seen to prowl in a wood of eighty acres, a short distance from my house; and night coming on before they were observed by the person who had charge of them, all attempts to recover them were in vain. No tidings being had of them, all attempts to rescue them proved fruitless; and as nothing was heard of them for several months, I of course concluded that they were either picked up by some persons who had stumbled upon them, or had been killed by foxes. About six months ago, however, I was riding through a large covert about half a mile from that into which my turkeys were seen to go, when a hen, apparently in a state of alarm, ran before my horse's feet, and disappeared in the bushes. It immediately occurred to me, as the color was the same, that she was one of the hens I had lost more than two years before. I have since caught her, and she is now in my yard. The other hen, which has young ones, was also caught, but on being given to a boy to hold, broke away from him again, and is still in the wood, with the cock-bird and the young ones. The one that I recovered had been sitting on eggs, as was evident from the state of her breast. Thus have these birds survived two winters, one a very severe one, in the woods, without either food or shelter, except that which nature provided for them."

That the turkey is susceptible of education is obvious from the following very interesting facts communicated to the author by Mr. L. Kennedy of Hartford. "A few years ago I purchased a pair of turkeys, kept them through the winter, and in the spring, instead of laying at home, they absconded. After hunting them up a number of times, sometimes finding them two miles off, I killed them. In the fall I procured another pair, and as the snow melted away they began to play at the same tricks; but by shutting them up I obtained a few eggs, and raised three young ones. When they were about half grown, I killed the old ones. One of the young ones was needed for 'thanksgiving,' and the remaining two I kept through the winter. About the time for the hen to commence laying I killed the cock, to prevent too much conversation and intimacy with distant flocks

That year I raised eight young ones, and they seemed to have made considerable progress towards becoming domesticated.

"Last year I kept over two hens and a male bird, and in the spring they both laid near the house, one of them under a bush in the garden, and the other in the barn. The one in the garden laid her third egg on the morning after the last snow that season, which was, I believe, the last of March, or fore part of April. I discovered her nest by the tracks in the snow. Supposing the first two eggs were injured by the cold, I left them in the nest, and removed the succeeding ones from day to day, until I had taken out sixteen, when she began to sit. I then removed the two eggs and placed the sixteen in the nest, adding also one from the nest of the other turkey. The other turkey sat on fourteen eggs and hatched out twelve; only one, however, proved to be rotten, the other was broken in consequence of the nest being too dishing. I did not remove the eggs from this nest as they were laid. I placed the old turkeys in coops near to each other, and consequently I cannot say how many of each brood died. The season was quite wet, so that I lost eleven and raised eighteen. I should have raised more probably, had I been situated so as to have let them run sooner, but as it was I succeeded much better than most of the farmers in this section. I give the young ones no food for the first twenty-four hours or longer, leaving them to peck the stones and dirt. After this I feed moderately with the curd of sour milk, never with clear meal, until they are several weeks old."

My turkeys, I am quite confident, not only know my voice but my person; and why not? "The ox knoweth his owner, and the ass his master's crib." This season they have not rambled to any distance, and usually taking care to be at home at meal times. But if all this is not sufficient to satisfy you of their being more domesticated than the "commonalty," let me add another circumstance which somewhat surprised me. The first of November I changed my residence, moving about thirty or forty rods from my former house, designing to remove the turkeys as soon as I had prepared a place to *confine* them. But the day after I removed, the turkeys followed one of the members of my family without his calling them at all, and came to the barn of their present home, when I fed them. At night they flew into an apple-tree in the garden, and have given me not the least trouble since. The tree where they formerly roosted is within sight, and yet they have never been into it, but have remained perfectly contented with their new location. I have nine of them now, all but one of which I design to put on my farm in the spring. The males which I killed the 16th of November, averaged about 10 pounds; and one of them weighed $10\frac{3}{4}$ pounds, all young ones.

"I think in an ordinary season, I can tenfold a flock of turkeys at an expense not exceeding twenty cents each."

As a further proof that turkeys may be made as tame and domestic as any other fowl, I will relate what occurred on my own premises. About two years ago I purchased a cock and two hen turkeys of the white variety. They had been hatched in the woods and suffered to run at large, and in the fall they selected some tall pines for their roosting place. On bringing them home and putting them in the yard with the other fowls, they refused to stay there, and would not roost in the house at first, preferring the top of the building, or a tall elm tree standing near. As the cold weather approached, and feeding them only in the house, they finally took to roost with the other fowls, and have remained there ever since.

They soon became very quarrelsome and would not suffer the other fowls to sit near them, pecking with their bills and throwing them from their roost. For which reason we attempted to keep them out of the yard, and removed them to the barn-yard, where there was a good shed for them to roost in. While the snow was deep they remained tolerably quiet, but as soon as the weather and snow would permit, they found their way to the yard again. We removed them the second and third time, feeding them plentifully, but all to no purpose; leave the poultry-yard they would not.

The following account of the propensity in a cock turkey for incubation was related by a correspondent of the Genesee Farmer. "The circumstance," says the writer, "I am about to relate, as far as I know, is not common, if it exists at all. I have been in the habit of rearing a good many domestic fowls, and among them have been rather partial to the turkey, particularly to fat ones about Christmas. Among a brood I once possessed, there was one male, who was a long-legged, gander-shanked fellow, of a most unique appearance. During the period of incubation, or as soon as one of the hens began to sit, which she, seeming to know the old gentleman's propensity, was very careful to manage in a very private and secret manner; he began to grow uneasy, and mounted the stumps and fences, watching for the appearance of the hen, and peering about to find the place of her concealment, which he usually discovered the first or second day; when he, by virtue of his authority as one of the lords of creation, immediately took possession of the nest, and from that time forward, till the period of hatching, went on with the regular process, when he brought off his brood, and duly carried them forward to maturity; when the hen, poor simple wife, was allowed to trudge along at a respectable distance, in the true after-honeymoon style.

"Although I am aware that certain other birds, male and female, alternately sit upon the nest during the period of incubation, yet I am not informed of any case where a male has shown such a decided passion and propensity, for the sedentary habit of hatching eggs; this he has performed for three years in succession, and being such a notable exhibition of pugnacious opposition to petticoat government, he became quite a favorite, and I intended to have kept him as an example to some of my neighbors, and as a *rara avis in terris*. But one night he came up missing, and whether he was sacrificed as a target at a Christmas gambol, or made one at Master Reynard's supper, or is even yet sitting on eggs that proved addled, I was never able to ascertain."

"The antipathy," says Mowbray, "which the turkey cock entertains for anything of a red color, is well known; and indeed will never be forgotten by myself, who, at about the age of eight years, having on a red waistcoat, was chased by two of them around a very extensive yard, to my most terrible affright and discomfiture."

It has been often repeated, that extreme difficulties occurred in raising the turkey; and that when, by dint of great pains, we had succeeded so far as to secure them from those accidents which threaten them till the time when the red color of the head shoots, the expenses they afterwards incurred to bring them to a desirable plumpness, exceeded the produce of the sale; this was sufficient to deter farmers from admitting this bird into their farm-yards, and they have been consequently deprived of a sure means of increasing the resources of the house, and at the same time of adding to the resources of the farm.

"It is important," says Dickson, "in breeding animals, to attend to their natural instinct as much as possible, and it is no doubt, from the neglect of this, that all the degeneracy and difficulty of rearing them which occurs, may arise."

In its native forests, the turkey is naturally a wandering and migratory bird, and hence it is unnatural to confine it to the narrow range of a poultry-yard; we speak from experience, as we found it, to our sorrow true, for one or two years. They have a strong disposition to wander, and will sometimes steal away a long distance from home, apparently wishing not to be observed.

Laborious efforts are not here required, but some care and a little patience. If attempts to raise turkeys have not been crowned with success, it is entirely owing to the unskilfulness and inexperience of those to whom they have been entrusted. And as long as it is persisted in thwarting the females when sitting, in opening the shells of the eggs in order to help the passage of the tardy chicks, in pressing them as soon as born, to make them eat against their will, in leaving them exposed to intense heat or cold damp; so long will their death be the undoubted consequence of such usage in the course of a month. It is less trouble to say that the bird is difficult to rear, than to acknowledge, at once, that negligence, unskilfulness, and barbarity, were the causes.

Delicate as they are supposed to be, they can find their living in the woods, and feeding on acorns, roots, berries, insects, and wild nettle seed. It is curious to observe with what adroitness and certainty they will pounce on the smallest bug or fly. The strong propensity of turkeys to perch in the open air, and on high places, is a sufficient reason for those who rear them, to attend carefully to this point. Scarcely, indeed, does the *red* appear, when the fowl shows his unconquerable desire to perch in the open air, though this cannot safely be permitted till they are two or three months old. Open sheds are consequently best suited to them, with roosting bars, fixed as high as convenient from the ground, which should be about three times as large as for common fowls.

Pairing.—Some writers say from six to eight hens may be allowed to one cock, while others assign from ten to twelve. " Your turkey-cock," says Gervase Markham, " should be a large and stout bird, proud, and majestical, for when he walketh dejected he is never good." And Mascall says, " he should be passing a year or two years old—three years is the most—and too much." According to Parmentier, both the cock and hen ought to have short legs, a full shape, and great vivacity and energy in all their actions; for breeding, it is peculiarly necessary that both should be well formed, and in healthy condition.

The plumpness or leanness of the hen, the climate or localities, will alone forward or retard her laying. By feeding and taking proper care of her in winter she will be disposed to lay earlier in spring, and to begin afresh at the end of the summer. Nature seems to have taken all the trouble on herself.

20

Laying.—In the wild state, about the middle of April, when the season is dry, the turkey-hens begin to look out for a place in which to deposit their eggs. In a domestic state, the time of laying is usually a month earlier than that of the wild turkey. It may readily be seen, indeed, when a hen is about to commence laying, by her vivacity, and also by her endeavors to secrete herself, and steal away from the observation of the keeper. She utters, besides, a peculiar note, indicative of her feelings; and when this has once been heard, it can never afterwards be mistaken.

The time of laying is almost invariably in the morning of every second day, though some hens will lay every day, till from fifteen to twenty eggs have been laid; in a wild state, more usually from ten to fifteen, according to the age of the bird; for a young bird, two or three years old, will lay nine, and larger eggs, than when only one year old.

During the time of laying, it is advisable to confine the cock, at least in the morning, when she is laying; otherwise if he finds her on the nest, he will ill-treat her, drive her away, and break her eggs.

It does not appear, from Audubon's account, that the wild turkey has usually more than one brood in the year, unless her eggs have been carried off or destroyed; and Buffon says, the tame turkey only lays once a year, which is wrong, for, in favorable circumstances, when well fed and taken care of, the hen turkey will lay a second time, towards the end of summer, sometimes sooner and sometimes later

In the second laying, there are rarely more than a dozen eggs; and in order to have the brood from these successful, more than ordinary care will be required.

Hatching.—A turkey-hen is one of the most steady and inveterate sitters of any known fowl. Before she has even completed her laying, she shows a wish to sit by unequivocal signs; she clucks like the common fowl, and continues in her nest till her breast becomes bare of feathers. While laying or sitting, she never moves, when an enemy passes in sight, unless she knows she has been discovered, but crouches lower until he has passed; hence the difficulty of finding them when laying abroad.

In the domestic state, the instinct of the turkey-hen is truly remarkable; her little artifices and tricks to conceal her eggs, and to deceive those who might try to discover her nest, appear almost dictated by reason; but what brings her back to the rank of a brute, is her manner of sitting; for even when her eggs are taken away, she will continue to sit on any substance whatever, even stones. It is, therefore, a matter of consequence that she be satisfied; for sitting without eggs would fatigue her more than natural hatching.

When turkey-hens have been left to themselves during their laying, and have chosen a nest at a small distance from the house, there is hardly anything to be done, for they will leave it with difficulty, and it is even prudent not to thwart them, for they generally hatch their own brood safely, and the younger ones are the stronger for it.

The turkey-hen sits from thirty to thirty-two days, and it is said will continue on the nest, even until starvation; and when hatched, she is not the most careful mother, nor is she a good provider, as she does not scratch for her young, like the hen, but leaves them to shift for themselves; but she is very alert to discover birds of prey and give timely notice As the young, at the moment of their birth, give no signs of seeking their food, and as they are not instructed in the least so to do it by the mother, it appears necessary to admit two or three eggs of the common fowl to those of the turkey-hen, ten days after sitting, so that the young ones may be hatched at the same time; as the common chickens peck and eat as soon as out of the shell, they become for turkey-chickens an example which they imitate, and which determines them to eat a few hours sooner, which is of some use.

The hen and brood must be housed during one month or six weeks, dependent upon the state of the weather. The scorching sun and the rain are above all hurtful to them; superfluous moisture, whether external or internal, is death to chickens, therefore, all slop victuals should be rigorously avoided. The utmost cleanliness is necessary, and a dry gravelled layer is most proper. High places exposed to the east or south are those which always agree best with the chickens, especially when they have a small separate yard.

Food.—The French, and all foreign writers, recommend for their first food, "bread crumbled and soaked in wine." The best food, however, which we have found for them, was eggs boiled hard and chopped fine; thick sour milk, boiled, which makes a thick curd; the whey is separated by putting it into a colander, or coarse sieve, and when cold, rubbed fine in the hands, and fed to them in small parcels, and often. Indian meal wet in the ordinary way is injurious to them, until they are several weeks old. Chives cut fine and mixed with their food, is eaten with great avidity. In case of the chicks appearing sickly and the feathers ruffled, indicating a chill from severity or change of weather, ground malt with a little barley-meal, is allowed, and by way of medicine, powdered caraway or coriander seeds. Boiled meat pulled into strings, in running after which the chicks have a salutary exercise. It will be borne in mind, that the above diet is beneficial for every other species of chicks, equally with the turkey.

Water should be given them in very shallow vessels, in which they cannot wet themselves, as this would be very injurious. In order to prevent the mother turkey from robbing the chicks of their food, they should be fed in a separate coop at such a distance from her as to be out of her reach.

Some recommend removing the chicks from the mother as they are hatched, but we are not of that opinion; nature seems to be the best guide, as they generally keep under the mother, as animal warmth is, without doubt, infinitely more necessary to them than food. It is well known, that birds on leaving their shells quit a warmth of sixty or seventy degrees, and that they often perish, sooner or later, on account of the difference of temperature through which they pass so suddenly; and being so excessively delicate, they should not be taken from warmth and repose; therefore in the beginning of his existence, the new-born chick remains under the wings of his mother, where he finds the warmth nearly equal to that he had in the egg; by removing him from this shelter, to handle and feed him, he passes too suddenly from heat to cold, from rest to exercise; and this sudden change, so hurtful to grown animals, becomes more especially so to the turkey-chicken, where natural delicacy and want of feathers render him more sensitive to the transitions.

When young, they should be kept in a warm and dry place; and when introduced into the open air, it should be by degrees, and choose the finest days. They should not be suffered to go abroad in the morning till the sun has dried the dew, and they should be shut up before the fall of the evening damps. On their return at evening, they should be fed, except in harvest time, when they have gathered enough in the fields.

The mother leads them with the same solicitude that the hen leads her chickens; she warms them under her wings with the same affection, and protects them with the same courage. It would seem that tenderness for her offspring, gives quickness to her sight; she discovers a bird of prey at a prodigious distance, when it is yet invisible to every other eye. As soon as she perceives her dreaded enemy, she vents her fears by a scream that spreads terror through the whole brood; each little turkey seeks refuge under a bush, or squats in the grass, and the mother keeps them in that situation by her cries, so long as danger is impending; but when her apprehensions are removed, she informs them by a different note, and calls them from their concealment to assemble around her.

Young turkeys are interesting, for they have different tones, and different inflections of voice, according to their age, their sex, and the various passions by which they are influenced; their pace is slow, and their flight tardy.

At two periods of their lives, turkeys are very apt to die; viz. about the third day after they are hatched, or when they throw out what is called the "red head," which they do at about six weeks old. It is a very critical period in the life of a turkey—much more so than the period of moulting; the food must, therefore, be increased, and rendered more nutritious by adding boiled eggs, wheaten flour, bruised hemp seed, or a few bruised beans.

Turkeys are the most tender and difficult to rear of any of our domestic fowls; but with due care and attention, which, rightly considered, in all things gives the least trouble, they may be produced or multiplied with little or no loss; and the same may be said with all truth of the rest of our domestic fowls, and animals in general, the losses and vexations annually deplored, arising almost entirely from ignorance and carelessness united hand in hand. Turkeys as well as geese, under a judicious system, may be rendered an object of a certain degree of consequence to the farmer.—*Mowbray.*

Of the fattening of turkeys.—It is only when the cold comes, and when turkeys are about six months old that they should be fed with better and more plentiful food in order to increase their size and plumpness for market. Indian corn, ground barley, wheat, also rice and other articles, used to fatten common fowls, are considered best for turkeys. Their weight, when well fattened and carried to market, should average twelve pounds; their living and dead weight is as eighteen to twelve pounds.

Cobbett says—"As to fattening turkeys, the best way is never to let them get poor. Barley meal, mixed with skimmed milk, and given to them fresh, will make them fat in a short time. Boiled potatoes mixed with Indian meal, will furnish a change of sweet food which they relish much, and of which they may eat as much as they can. As with others, the food of this bird must be kept clean, and the utmost care taken not to give them on the morrow the remains of the mixture of the preceding day; because if the weather is warm, it will sour, which might displease them."

In some sections, in order that they may all get fat more *expeditiously*, some professional poulterers cram their turkeys, which, with the barbarous practice of depriving them of sight, and light, and motion, by confining them in narrow enclosures, is so revolting to humanity, that it is to be hoped that the horrid custom is but seldom practised, in this country, at least. The very idea is enough to disgust and cloy the appetite of the most consummate epicure!

There are said to be some advantages attending this mode of feeding turkeys; but it is, to say the least, unnatural and cruel; and therefore, fattening in freedom, and as they naturally choose, is a more certain way of procuring pure, healthy birds, free from all plethoric disease.

We give the following, which was published in the National Recorder, more for its singularity, than in the belief of its usefulness. It reminds us of the account published in the "American Swine Breeder," relative to using "hay tea" for hogs. A friend, who tried the experiment, was so unfortunate as to lose a large number of very valuable swine before he could be convinced of its worthlessness, as food for hogs, without an admixture of other and more substantial food. For accommodation sake, we will believe it to be true, but we advise all those who may read this to be cautious how they try experiments so unnatural.

The writer alluded to, remarks,—"In the winter of 1818-9, a gentleman in this city made the following experiment. He placed a turkey in an enclosure about four feet long, two feet wide, and three or four feet high. He excluded as much light as he could without preventing a circulation of air, and fed the turkey with soft brick broken to pieces, with charcoal also broken, and with ten grains of corn per day. Fresh water was daily supplied. The box or coop in which the turkey was placed, he always locked up with his own hands, and is perfectly confident that no person interrupted the experiment. At the end of one month, he invited a number of his neighbors; among others, two physicians. The turkey, now very large and heavy, was killed and opened by the physicians, and was found to be filled up with fat. The gizzard and entrails were dissected, and nothing was found but a residuum of charcoal and brick. To conclude the examination satisfactorily, the turkey was eaten and found to be good.

"Last winter he again repeated the experiment with the same success.

"The circumstance which induced him to make the experiment, is a very curious one. One of his neighbors informed him, that being driven from the city by the fever of 1793, his family recollected that some fowls that had lived in a kind of loft over his work-shop, had been forgotten in the hurry of their removal, and would certainly be starved. They were absent six or eight weeks, and on the retiring of the pestilence returned. To their great astonishment, the fowls were not only alive, but very fat, although there was *nothing but charcoal and shavings* that they could have eaten, and some water that had been left in the trough of a grindstone, had supplied them with drink."

Since the publication of the fourth edition, the author has received the following very interesting and valuable suggestions from Col. Wade Hampton, of South Carolina.

"I have read your treatise on poultry with much interest, and have derived from its perusal very many valuable suggestions.

"On the subject of rearing turkeys, I venture to make you a suggestion or two. As soon as they are removed from the nest, immerse them in a strong decoction of tobacco, taking care to prevent the fluid from entering the mouth or eye of the chick, and repeat the operation whenever they appear to droop. Corn-bread, soaked in pepper tea, is the best diet for them, after they are two or three days old, that I have ever tried. They are particularly liable to chills, which nothing so effectually cures as the pepper. A tablespoonful of Cayenne pepper to a quart of boiling water is about the rate at which it is used."

Fig. 53.

GUINEA HEN.

The Guinea Fowl, as its name imports, is a native of Africa, and is supposed to have been introduced soon after the Europeans had visited the western coast of Africa, in their voyages to India by the Cape of Good Hope. And they not only have diffused these birds through Europe, but transported them into America; and the guinea hens have suffered various alterations in their qualities, from the influence of climates.

" Charlevoix pretends that there is at St. Domingo a species smaller than the ordinary sort. But these are probably the Chestnut Pintador, bred from such as were introduced by the Castilians soon after the conquest of the island. Those having become wild, as it were, naturalized in the country, have experienced the baneful influence of that climate."

The plumage of the guinea fowl, though not decorated with rich and dazzling colors, is remarkably beautiful. It is of a blackish-grey ground, sprinkled with considerable regularity with white rounded speckles, resembling pearls. Such at least was the plumage in its native climate; but since it has been carried into other countries, it has assumed more of the white. Thus the Pintados at Jamaica and St. Domingo are white on the breast; and Edwards mentions some that are entirely white. We have also heard of one in this country that was nearly black. Its short wings and pendulous tail, like that of the partridge, which, joined to the arrangement of its feathers, make it look as if it were hunch-backed; but this appearance is false in reality, as no vestige remains when dressed.

The size is that of an ordinary hen, but the shape is like that of the partridge; hence Brun called it the Newfoundland partridge. But it is of a taller form, and its neck longer, and more slender near the arch.

The guinea hen is an exceeding noisy bird, and for this reason Brown has termed it "*Gallus clamorous.*" Its cry is sharp, and by its continuance, becomes so troublesome, that though the flesh is very delicate, and much superior to that of ordinary poultry, most of the American farmers have given over breeding it.

The guinea hen is a lively, restless, and turbulent bird, that dislikes to remain in the same place, and contrives to become master of the poultry-yard. It can intimidate even the turkeys; for though much smaller in size, it gains the ascendency over them by the mere dint of petulance.

The guinea hen is one of those birds which by wallowing in the dust, rid themselves of vermin. They also scrape the ground, like common hens, and roam in numerous flocks. Bodies of two or three hundred together are sometimes seen in the Isle of May; and the inhabitants hunt them with a greyhound, and without other weapons than sticks.—*Dampier.*

The guinea hen lays and hatches like the ordinary hen; but its fecundity appears to be not the same in different climates, or at least that this is much greater in the domestic condition, where food is more abundant than in the wild state, which affords but a scanty subsistence. "I have been informed," says Buffon, "that it is wild in the Isle of France, and there lays ten or twelve eggs on the ground, in the woods; whereas, those that are domestic in St. Domingo, seek the hedges and bushes to deposit their eggs, laying 100 or 150, provided that one be left constantly in the nest." These eggs are smaller in proportion than those of an ordinary hen, and their shell is much harder.

The young Pintados are very tender, and being natives of the burning climate of Africa, are with difficulty reared in our northern climate. They feed on grasshoppers, worms, insects, and all kinds of grain, particularly millet.

"The Pintado cock," says Dampier, "breeds also

with the common hen, but is a kind of artificial union which requires attention to bring about. They must be bred together from their infancy, and the hybridous intercourse gives birth to a bastard progeny, and of an imperfect structure, and disavowed as it were by nature. Their eggs are destitute of the prolific power, and the race is extinguished in the death of the individuals."

Mr. Pennant makes it appear that the Pintados had been early introduced into Britain, at least prior to the year 1277. But they seem to have been much neglected, on account of the difficulty of rearing them; for they occur not in the ancient bills of fare. —*Buffon.*

The guinea hen has a great propensity for rambling; very restless and shy, laying abroad, and giving no indications, like the hen, of laying. Their eggs are excellent, and keep well. Their flesh, when young and fat, is very superior, dark-colored, and in flavor is said to resemble our partridge.

By their continued cries and watchful nature they are useful in protecting the other poultry from the hovering hawks—for which reason, if no other, a few should always be kept in the poultry-yard.

CHAPTER X.

AQUATIC FOWLS.

THE SWAN—ornamental—one at Fairmount—antiquity—longevity — varieties —Polish —Hooper— Bewick's—tame—black —feed—sitting—incubation—found in North America—cannot be kept north of Philadelphia—notice of a pair in New York—interesting account of a " weatherwise" Swan.

THE GOOSE—species —twenty varieties enumerated—famous for having warned the Romans of the enemy's approach—ponds, &c., not necessary—longevity—stimulant qualities of their flesh—pairing—selection—signs of laying—time of laying—early laying accelerates fecundity—will live on grass—good foragers—feed—can be raised at profit—estimate of profit—profits of a flock owned by two females.

WILD GOOSE—characteristics—flight—domesticated, not a species—will mate with domestic geese—produce mules—good qualities—great size—singular account of taking when their wings were covered with ice.

BREMEN—character—first introduction—premium awarded Col. Jaques's stock.

AFRICAN—great size—Buffon's description—Brisson's account—somewhat similar to Poland geese.

CHINESE —description—color of plumage—early laying —size—some perfectly white—attempt to domesticate the Brant.

HABITATION—selection of situation—not good to keep geese with other poultry—incubation—time—management of the goslings —pasture—solid food best—must not confine them.

DISEASES—prevention better than cure—diarrhœa—insects afflict them.

PLUCKING—number of times to be plucked—clipping recommended.

FATTENING—best time—food—French method —liver a great dainty with the Romans—singular trait of attachment to man—object in keeping.

Fig. 54.

THE WHITE SWAN.

At the head of this class of birds may justly be placed the stately and majestic *Swan.* Next to the peacock on the land the swan is the most noble and elegant fowl on the water. Though they are seldom found on any farms in this country, and are not in request as food, they are well worth the notice of every one having a pond or an enclosed part of a creek, to enliven and beautify the scenery by a small family of swans.

Every person that has visited Fairmount, Philadelphia, must recollect the beautiful swan that floats silently and majestically on the bosom of the little pond at its base. Its beautiful plumage of pure white, black legs, feet, and bill; its beautiful curved neck and graceful movements, could not escape the notice of any one.

The antiquity of this stately bird, the "silent swan," is conspicuous in the pages of history and poetry. The prototype of the domesticated breed has been probably lost in the lapse of time, "since the wild swans," says Mowbray, "of all countries, differ essentially both in plumage and organic structure from the tame. The longevity of the swan seems to equal, if not exceed, that of any other animal, as it is said to live three centuries."

"The goose, the swan, and the eagle," says Boswell, "are well known to be the longest lived birds. Of the former, it is comparatively easy to discover the precise age; of the third, from its very nature, it is clearly impossible; and of the second, from its temporary overpowering propensity to change situations, it is very difficult. The place of an old swan may be supplied by a younger one, and may still from their similarity be considered the same."

Besides the tame swan, there are said to be three European varieties. Of these, one has been recently characterized; it is allied to the tame swan, but instead of the legs, toes, and web being black, as in the latter, they are of a pale, ashy grey. The cygnets are white. Mr. Yarrell, the first discoverer of this species (of which several individuals are living, and have bred in the garden of the Zoological Society), observes, that "this species had been known to him for some years past, as an article of commerce among

London dealers in birds, who receive it from the Baltic, and distinguished it by the name of the Polish swan. In several instances, these swans had produced young in this country, and the cygnets when hatched were pure white, and did not at any age assume the brown color borne for the first two years by the young of all the other species of swan." —*Proc. of Zool. Soc.*, 1838.

During the severe winter of 1837-8, flocks of the swan were seen pursuing a southern course along the line of our northeast coast, from Scotland to the mouth of the Thames, and several specimens were obtained. One flock of thirty, and several smaller flocks, were seen on the Medway. The skull of this species differs in certain points from that of the tame swan, according to Peleren, who has published a paper on the subject in the Magazine of Natural History, April, 1839. Of the two remaining swans, one is the wild swan, Hooper, or Whistling Swan, a native of the whole northern hemisphere, breeding on the borders of the arctic circle, and migrating southward in winter. In America, the emigration of this swan is bounded by Hudson's Bay on the North, and extends southwards as far as Louisiana, and the Carolinas. It extends its winter visits in Europe and Asia as far as the warmer latitudes, and passes into Egypt. The wind-pipe of this swan is remarkable for a loop which passes into the substance of the keel of the breast-bone.—*Penny Mag.*

The last European species is Bewick's swan,

which has been confounded with the Hooper, bu which, as Mr. Yarrell has demonstrated, is a distinct species. Like the preceding, it is a native of the high northern regions, migrating south in winter. Its windpipe is of smaller calibre than that of the Hooper, and passes far more deeply into the keel of the breast-bone.

" The tame swan," says Dickson, " is very different from the wild swan which are sometimes seen in this country (England), though by no means common."

BLACK SWAN.

This bird is found in large flocks in Van Dieman's Land, and on the western coast of New Holland or Australia, and is sufficiently tamed to breed in this country. The black swan was first found, at Swan, or Black Swan River, by a Dutch voyager, who, in 1697, sailed forty or fifty miles up the river in his boat.

When Capt. Flanders, an excellent sailor of late years, first explored the same coast, he found black swans in immense flocks, in the openings both of the rivers Tamer and Derwent. Of these flocks, he says, " From one-fifth to one-tenth of them were unable to fly ; they cannot dive, but have a method of plunging so deep in the water, as to render their bodies nearly invisible, and thus frequently avoid detection. In chase, their plan was to gain the wind upon our little boat, and they generally succeeded

Fig. 55.

BLACK SWAN.

when the breeze was strong, and sometimes escaped from our shot also. It is black in every part, except on a few of the quill-feathers, which are white. The bill is bright red above, and greyish-white beneath.—*Child's Natural History of Birds.*

The black swan of Australia is sometimes reared here (England), and may, in time, come to be bred more extensively. It requires little different manage ment from the common swan.—*Dick.*

" The black swans of New Holland," says Mowbray, " I have not hitherto had the opportunity of seeing. They were introduced in this country some years since, but I believe the number bred or remaining is very small. They are said to degenerate here as to size, yet the imported individuals, it seems, were no larger than our indigenous breed. There is said by naturalists, to be some disparity between the wild and tame black swan, in respect to the bill and organization of the bones. Hence probably they form different species of the same genus."

" The tame or mute swan," says a writer in the Penny Magazine, " is abundant on the Thames, each pair having their exclusive range or district, at least during the breeding season. The nest, in the formation of which both male and female labor, is made on the banks, among reeds or oziers, on one of the ozier islands. It consists of a mass of sticks, or twigs, raised sufficiently high to prevent its being overflowed by any rise of the water."

The swan feeds like the goose, and has the same familiarity with its keepers, kindly and eagerly receiving bread which is offered, although it is a bird of courage equal to its apparent pride, and both the male and female labor hard in forming a nest of water plants, long grass and sticks, generally in some retired spot; and they are then very dangerous to approach, their size and strength enabling them to break a man's limb with a stroke of their wing. The hen begins to lay in February, producing an egg every other day, until she has deposited seven or eight, on which she sits six weeks. Buffon says it is nearly two months before the young are excluded. Swans' eggs are much larger than those of the goose, white, and with a hard and somewhat tuberous shell. The cygnets are ash-colored when they first quit the shell, and for some time after; indeed, they do not change their color, nor begin to moult their plumage, until twelve weeks old, nor assume their perfect glossy whiteness, until advanced in their second year.

Swans cannot be made to thrive without abundance of water to swim in, and clear water is to be preferred to that which is muddy.

"The swan is found in various parts of North America. Here this noble bird is seen floating near the shore in flocks of some two or three hundred, white as the driven snow, and from time to time emitting fine sonorous and occasionally melodious notes so loud that they may be heard on a still evening, two or three miles. There are two kinds, so called from their respective notes; the one the trumpeter, and the other the whooper; the former is the largest. These birds are sagacious and wary, and depend more on sight than on the sense of smell."

It is doubtful whether the swan could be kept through the winter north of Philadelphia. Our winters are too severe for them, and it is necessary for them to have water to resort to sufficient for them to swim in, in winter. Mr. Prentice of Albany imported a pair from England, a few years since, and placed them in artificial ponds supplied from springs, but they did not do well and finally died.

Since writing the foregoing, we noticed a pair of these splendid birds sporting in the basin in the

Bowling Green, at the foot of Broadway, New York. In November last, when we saw them, they had not recovered from their confinement in cages, and looked rather rough; still they added much to the beauty of the scene.

A very interesting account of "a weather-wise Swan," we find in an English paper, which we transcribe. "This swan, the property of Lord Braybrooke, which was eighteen or nineteen years old, had brought up many broods, and was highly valued by the neighbors. She exhibited, some eight or nine days past, one of the most remarkable instances of the power of instinct ever recorded. She was sitting on four or five eggs, and was observed to be very busy in collecting weeds, grass, &c., to raise her nest. A farming man was ordered to take down half a load of haulm, with which she most industriously raised her nest and the eggs two feet and a half. That very night there came down a tremendous fall of rain, which flooded all the malt shops and did great damage. Man made no preparation—the bird did. Instinct prevailed over reason—her eggs were above and only just above the water."

THE GOOSE.

"In the species of the *Goose*," says Boswell, "*properly so called*, nature knows but one race. The industry of man has created another, larger, whose shape, color, as well as nature, have undergone those modifications which are to be observed in all animals that have for a long time been domesticated."

Fig. 56.

THE COMMON GOOSE.

Dickson says there is only one sort of the common goose, though there are several varieties of the tame goose. The wild goose is of a brownish-ash color, the individual feathers being lighter on the edge; the belly is snow-white. The tame varieties are of various colors, chiefly white, and all the various shades of grey.

Main enumerates twenty-two varieties, viz.: Antarctic Goose, spur-winged Goose, Bering Goose, black-backed Goose, Canada Goose, barrel-headed Goose, rufus-necked Goose, half-webbed Goose, Egyptian Goose, Esquimaux Goose, Chinese Goose, snow Goose, Kusarka Goose, harvest Goose, mountain Goose, painted Goose, laughing Goose, great Goose, Magellanic Goose, grey-headed Coromandel Goose, variegated Goose.

22

The goose, famous for the services it rendered the Romans, by warning them of the enemies' approach under the walls of the Capitol, is next in size to the swan, of the aquatic species, is considerable of an object in rural economy, and, in favorable situations, is considered of some importance, on account of its large size, product of feathers, and market value, and is found on most farms where there is a stream or pond of water near, for without them they will not thrive and do well. It is best to have a lot enclosed for them to pasture in, as they must not be suffered to graze on the land devoted to other stock, as cattle will not eat after them. Some farmers contend that their dung is poison to land, and grass will not flourish where they are pastured, probably in consequence of their being over-dosed.

M. Parmentier, and some other writers, assert that the vicinity of ponds and rivers is not absolutely necessary to the successful rearing of geese, for in districts destitute of these advantages, a small reservoir where they can swim will be sufficient.

Of all the stock brought up on farms, the goose lives to the greatest age; there are records of some attaining to a century. In 1824 there was a goose living in the possession of a Mr. Hewison (England) which was then upwards of one hundred years old; it had been throughout all time in the constant possession of Mr. H.'s forefathers and himself; and on quitting his farm he would not suffer it to be sold with the rest of the stock, but made a present of it to the incoming tenant, that the venerable fowl might terminate its career on the spot where its useful and long life had been thus far spent.—*Dickson.*

There was also lately a goose on a farm in Scotland of the clearly ascertained age of eighty-one years, still healthy and vigorous; she was killed while sitting on her eggs by a sow. It was supposed she might still have lived many years, and her fecundity appeared to be permanent. Other geese have proved fertile at seventy years.

"The whole anserine or goose tribe," says Mowbray, "of which there is a great variety, are held to afford a food highly stimulant, of a viscous quality, and of a putrescent tendency. The flesh of the tame goose is more tender than that of the wild, but generally it is a diet best adapted to good stomachs and powerful digestion, and should be sparingly used by the sedentary and weak, or by persons subject to cutaneous diseases."

PAIRING.

In most works on poultry, as well as in general practice, it is usual to assign one gander to five or six geese. But from our experience, and it has been ascertained by M. St. Genis, that geese pair like pigeons, partridges, and many other birds; and in order to have success in the rearing of goslings, it is better to have as many ganders as geese, at any rate not more than two geese to one gander. If the number of ganders do not exceed the geese, including the common father, no disturbance or disputes occur, the pairing takes place without any noise, and no doubt by mutual choice. Where the number of geese much exceeds the number of ganders, many of their eggs are unfruitful—at least such has been our experience.

In selecting ganders they should be of a large size, with a lively eye and an active gait; the geese, if of the common kind, party-colored, or the lightest colored, as they produce the best of feathers, with the exception of the Bremen, which are always white.

Whenever it is perceived that a goose wants to lay, which is known by her carrying straw in her bill, an essential precaution is to coop them up under their shed, where nests, made of fine soft straw, have been prepared; as soon as the first egg is laid, they continue to lay successively in the same place.

If the weather will permit, it will be well to hasten the laying of geese, in order to have early goslings, that they may be strong and come to their utmost size when the time comes for selling them.

Dickson observes that "when well fed, geese will lay thrice a year, from five to twelve eggs each time, and some more, that is, when they are left to their own way; but if the eggs be carefully removed as soon as laid, a goose may be made, by abundant feeding, to lay from twenty to fifty eggs without intermitting. They begin to lay early in spring, usually in March."

Mowbray observes, " The earliness and warmth of

spring are the general cause of the early laying of geese, which, of consequence, since there may be time for two broods within the season, not however a common occurrence, and which happening successively for two or three seasons, has occasioned some persons formerly to set a high price on their stock, as if of a peculiar and more valuable breed than the common. The method, however, to attain this advantage is to feed breeding geese high throughout the winter with solid corn, and on the commencement of the breeding season, to allow them boiled barley, malt, fresh grains, and fine pollard mixed up with ale or other stimulants. Instances are said to have occurred of a goose laying upwards of one hundred eggs within the year."

The fecundity of the goose is well known; she lays many eggs and of a large size; they are not so good as the hen's for the table, but can be used in pastry to advantage; being large, one will supply the place of three hen's eggs.

A Mr. Holmes had a goose in his possession which, within the year, laid seventy eggs; twenty-six at the usual time of incubation, from which she hatched and brought up seventeen fine goslings. She began to lay again at the end of harvest, and continued to lay every other day to the end of the year, and remained in high condition.

A goose of the small Chinese variety, now in possession of the author, commenced laying in November and continued until in the winter, when the number of her eggs amounted to forty-five. She commenced again in April and laid another litter of over thirty eggs, one of which for cooking is full equal to three hen's eggs.

The profit to be derived from geese feathers is not anywhere to be neglected; it is an important article and always commands a fair price. An acquaintance, who is very particular in keeping the feathers clean, finds a ready market at from 50 to 60 cents per pound. A goose will yield from 15 to 17 ounces in a season.

Geese will live entirely on grass, but will pay for grain if rightly managed. Oats are the best food, and cabbage, turnips, parsnips and carrots are good for them. "Grass," says an old writer, "they must necessarily have, and the poorest and that which is most useless, is the best, as that which is moorish and unsavory for cattle." Columilla also advises to pasture geese in marshy or moist grounds, and to sow for them vetches, tares, or clover.

Did geese require to be always *fed* with grain in the poultry-yard, they would be found to *cost more* than they would fetch. Though they consume a vast quantity of food, they are easily fed and are excellent foragers. Cobbett says, "the refuse of the cabbage of a market garden would maintain a great many geese at a small expense, but it is doubtful whether they would keep long in good health when fed on cabbage and lettuce, as it is apt to render their bowels too open and bring on scouring, unless attended with boiled or steamed potatoes, or ship stuffs, Indian meal mixed up with boiled potatoes, carrots or turnips. It is, however, a good practice always to feed a moderate quantity of any solid food, such as corn or oats, to the flocks of store geese, mornings and evenings, before going out, and on their return in the evening more especially; and by such full keeping they will always be in good flesh and attain a large size; the young ones will also be forwarded and valuable for breeding stock.

Geese managed on the foregoing mode will be speedily fattened at six or eight weeks old, or after the run of the grain stubbles. With clean and renewed beds of straw, plenty of clean water, and fed upon oats, corn or peas ground, or ship-stuffs mixed with skimmed milk, geese will fatten pleasantly and speedily. Very little greens of any kind should be given them when fattening, as being too laxative, whence their flesh will prove less substantial and of inferior flavor.

It is stated in the Farmer's Gazette that geese can be raised, in a proper situation, at a profit far greater than almost any other stock. But to do this, more attention is required than is usually bestowed on their keeping and management.

"But let us make an estimate of the profit of ten old geese, in the manner they are generally kept by most farmers. We will suppose that the goose-keepers (for there are those who are not farmers) commence operations by purchasing ten geese in the spring before they begin to lay, at one dollar each, which is a quarter more than they can be frequently obtained for. Eight of the ten geese (for two should be ganders) will have on an average ten goslings each, but allowing one half for paper calculation, and probably less through the season, it will leave us with a flock of fifty, old and young, worth when dressed for the market not a dollar,—the original cost,—but half this sum, and you have twenty-five dollars. In addition to this, every old goose will yield one pound of feathers, and every young one three-fourths of a pound, making in all forty pounds, which added to the twenty-five gives us forty dollars. We say nett profit, for there is not one goose-keeper in ten that feeds his geese, either old or young, after the grass has started in the spring, until fattening time in the fall: and then the large quills will more than pay for their extra food."

The above calculation is made having reference to the usual mode of managing this fowl, which is no management at all. Because, in the first place, they have generally no place to obtain their food, but on the open commons, except such as they too often steal from meadows, to the great injury of the standing grass, to the feelings of its owner, and very frequently putting their own necks in jeopardy.

But, on the other hand, if the owner will provide a good, warm and dry house for the accommodation of his geese while laying and hatching, and attached to this a pasture, where they may at all times have access to green grass and a small stream or pond of water, with due attention, and the right bird, which in our opinion are the Bremen, and our word for it, with only ordinary good luck, he will receive more than ordinary profit on the care bestowed and capital invested. Bremen geese are larger than the common geese, are always white, and yield on an average, from one to three ounces more feathers, and these of a better quality (having more down attached to them) than those of the common brown goose.

A writer in the Maine Farmer says, "I once knew a couple of industrious sisters who lived near a never failing brook or stream, in Massachusetts, who kept generally through the winter thirty geese, male and female. They had erected some suitable but not costly sheds, in which they had apartments for them to lay, sit and hatch. Their food in the winter was meal of various kinds, to some extent, but principally apples and roots. In the summer they had a pasture enclosed with a stone wall or board fence, which embraced the water. They kept their wings so clipped that they could not fly over such a fence. Their owners well knew (what we all know) that live geese feathers are a cash article at a fair price. They picked off their feathers three times in the season. Those thirty geese wintered, would raise seventy-five goslings or young geese, and of course they had that number to dispose of every fall or in the beginning of winter, when they are sent to market, and again picked, making four times they obtained feathers from those they wintered, and twice from the young ones that they had killed.

"I tell the story to induce some family, sisters or brothers, fathers or mothers, situate near some never failing brook of water, to go and do likewise. Those remote from water cannot be benefited by the history, yet their friends may; but if I can by this account cause one family to partake of the benefits of the business, I shall be satisfied. Many families there are, in all our towns, so situated that they may make the raising of geese a profitable business, yet, perhaps, have never thought of their privileges. It is known that we must import most of our feathers; and is it necessary to send abroad for an article so easily produced among us? Those who calculate to commence the business must prepare for it the ensuing fall, and not kill their geese. No one will object to the keeping of even more than thirty geese, if an enclosure is made sufficient to keep them at home and out of mischief."

VARIETIES—WILD GOOSE.

This bird is called by all the foreign writers the Canada Goose. It is not uncommon in a state of domestication, and is abundant as a wild bird, breeding in the arctic regions, and speeding south on the approach of winter; and its migrations north are the sure signs of returning spring. It is now to be classed among our domestic water-fowl, and is kept as an ornament to our ponds and sheets of water.

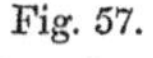

Fig. 57.

WILD GOOSE.

Wilson furnishes us with the following beautiful account of this bird: "I have never visited any quarter of the country where the inhabitants are not familiarly acquainted with the passing and repassing of the wild goose. The general opinion *here* is, that they are on their way to the lakes to breed; but the inhabitants on the confines of the great lakes are equally ignorant with ourselves of the particular breeding-places of these birds. *There* their journey north is but commencing, and how far it extends it is impossible for us at present to ascertain. They were seen by Hearne in large flocks within the arctic circle, and were then pursuing their way still farther north. They have also been seen on the dreary coast of

Spitzbergen, feeding on the water's edge. It is highly probable that they extend their migrations under the very pole itself, amid the silent desolation of unknown countries, shut out from the eye of man by everlasting barriers of ice. That such places abound with suitable food, we cannot for a moment doubt.

"The flight of the wild goose is heavy and laborious, generally in a straight line, or in two lines approximating to a point. In both cases, the van is led by an old gander, who every now and then pipes his well known *honk*, as if to ask how they come on; and the *honk* of all's well is generally returned by some of the party. When bewildered in foggy weather, they appear sometimes to be in great distress, flying about in an irregular manner, making a great clamor. On these occasions, should they alight on the ground, as they sometimes do, they meet with speedy death and destruction. The autumnal flight lasts from the middle of August to the middle of November; the vernal flight from the middle of April to the middle of May."

Wounded geese have frequently been domesticated and readily pair with tame geese, but their progeny are mules and will not breed. On the approach of spring, however, they discover symptoms of uneasiness, frequently looking up in the air and attempting to go off. Some, whose wings have been clipped, have travelled on foot in a northerly direction, and have been found at a distance of several miles from home. They hail every flock that passes over head, and the salute is sure to be returned by the voyagers, who are only prevented from alighting by the presence and habitations of man. The gunners sometimes take one or two of these domesticated geese with them to those places over which the wild ones are accustomed to fly; concealing themselves, they wait for a flight, which is no sooner observed by the decoy goose, than they begin calling aloud until the flock approaches so near that the gunners are enabled to make great havoc among them with musket shot.

Dickson says it is a variety, and not a distinct species, which from our experience we are led to doubt. We had a wild gander that was wounded in the wing, which paired with a common domestic goose, and we bred from them for more than ten years, but the produce were not fruitful, although they laid eggs. They never showed any disposition to pair or mate with either the wild or domestic goose. The hybrids partake largely of the wild habits, and if their wings are not clipped spring and fall, more particularly in the spring, they are very apt to fly away and not return. The author has lost two pair in that way; one pair, after whirling about the premises for a short time, bent their course towards the river, and alighted about three miles below Albany, where they were taken for wild geese and shot. The other two left in the winter, and after hovering about the neighborhood for two or three days, were seen to rise high in the air and direct their course towards the river, which was the last we ever heard of them. The old gander was finally shot in a small mill-pond near the house, by one of those lawless loafers who encroach on our premises much against our will. If he was shot to eat, we do not envy them their repast, for he was over thirteen years old to our certain knowledge.

Wild geese are regarded by those who have kept them, nearly as good and profitable as the domestic goose, which it exceeds in size, and especially in the quantity and quality of feathers; even the half-bloods show a great superiority in that respect. We have sold the young half-bloods, when fat, for $1 50 to $2 per head.

In Willoughby's Ornithology we find the following striking anecdote: "The following account of a Canada goose is so extraordinary, that I am aware it would with difficulty gain credit, were not a whole parish able to vouch for the truth of it. The Canada geese are not fond of a poultry-yard, but are rather of a rambling disposition. One of these birds, however, was observed to attach itself in the strongest and most affectionate manner to the house-dog, and would never quit the kennel except for the purpose of feeding, when it would return again immediately. It always sat by the dog, but never presumed to go into the kennel, except in rainy weather. Whenever the dog barked, the goose would cackle and run at the person she supposed the dog barked at, and try to bite him by the heels. Sometimes she would attempt to feed with the dog; but this the dog, who treated his faithful companion rather with indifference, would not permit. This bird would not go to roost with the others at night, unless driven by main force; and when in the morning she was turned into the field, she would never stir from the yard gate, but sit there the whole day in sight of the dog. At last orders were given that she should be no longer molested, but suffered to accompany it as she liked. Being thus left to herself, she ran about the yard with him all the night; and, what is particularly extraordinary, and can be attested by the whole parish, whenever the dog went out of the yard and ran into the village, the goose always accompanied him, contriving to keep up with him by the assistance of her wings; and in this way of running and flying, followed him all over the parish. This extraordinary affection of the goose towards the dog, which continued to his death, two years after it was first observed, is supposed to have originated from his having accidentally saved her from a fox in the very moment of distress. While the dog was ill, the goose never quit him day or night, not even to feed; and it was apprehended she would have been starved to death had not orders been given for a pan of corn to be set every day close to the kennel. At this time the goose generally sat in the kennel, and would not suffer any one to approach, except the person who brought the dog's or her own food. The end of this faithful bird was melancholy; for when the dog died, she would still keep possession of the kennel; and a new house-dog being introduced, which in size and color resembled the one lately lost, the poor goose was unhappily deceived, and going into the kennel as usual, the new inhabitant seized her by the throat and killed her."

The following singular trait of a sincere attachment of a goose to man, is related in Mowbray. "In March of the present year, 1829, Mr. Burnett had a goose nearly a year old, that formed so strong an attachment to him as to follow about through the crowd and bustle of the high street. It would attend him to the hair-dresser's shop, and patiently wait until he was shaved, after which, accompanying him to another shop of another person, proceeding thence home with him, cheek by jowl. This affectionate bird never fails to recognize its master under whatever change of dress; knowing also his voice, though not seeing him; and no sooner does he speak than it responds to him in its own unintelligible dialect. Had Butler been aware of a faculty like the above in the goose, he probably would not have be-rhymed it to the following purport:—

> "Art has no enemies
> Next the ignorant, but owls and geese."

The following we take from one of our newspapers, and if true, it exceeds anything of the Goose tribe we have ever heard of. "A gentleman of Brookfield, Conn., lately shot on Still River, eight wild geese at four successive shots, and wounded a ninth. The largest goose killed weighed near forty pounds."

Another wild-goose story we take from the Laparte Whig: "A most singular advantage has been taken of these wild fowl (geese) on the prairies in this country. We understand that during the late rain and sleet storm, large flocks of wild geese were so completely frozen over that their wings became useless appendages, and they were compelled to 'take to their heels' for their only chance of escape from the eager pursuer, and during this helpless and forlorn condition, large flocks of them were captured. The old saying is, 'when the sky falls we can catch larks,' but a new and more probable one is, *when the sleet falls we can catch geese.*"

BREMEN GOOSE.

This is a larage and beautiful bird, perfectly white, both male and female, with orange-colored legs and bill, and makes a beautiful appearance on a sheet of water.

They were first introduced into this country by Mr. James Sisson, of Warren, R. I., who gives the following account of them in the New England Farmer. "In the fall of 1826 I imported from Bremen (north of Germany) three full-blooded perfectly white geese. I have sold their progeny for three successive seasons; the first year at $15 per pair; the two successive years at $12. Their properties are peculiar; they lay in February; sit and hatch with more certainty than common barn-yard geese; will weigh nearly, and in some instances quite, twice the weight, have double the quantity of feathers; never fly, and are all of a beautiful snowy whiteness. I have sold them over the interior of New York, two or three pairs in Virginia, as many in Baltimore, North Carolina and Connecticut, and in several towns in the vicinity of Boston. I have one flock, half-blooded, that weigh on an average, when fatted, thirteen to fifteen pounds; the full-blood weigh twenty pounds."

Mr. Sisson received a premium from the "Rhode Island Society for the Encouragement of Domestic Industry," for the exhibition of geese of this breed. They are said to possess the following advantages over the other varieties of their species: they grow to a great size, may be fattened with less food, and their flesh of superior quality.

They may now be found in almost any part of the country; but the largest and finest flocks we have ever seen were in the possession of Col. Jaques, of Charlestown, Mass. We saw, when attending his auction in the winter of 1842, two hanging in his piazza dressed and marked "18 lbs." They were very fat, and said to have been goslings. We have a small breeding flock of them, but they are not equal in size to the above. Our goslings, when six or seven months old, with common feed, averaged when dressed from ten to twelve pounds.

Fig 58

GUINEA, OR AFRICAN GOOSE.

This is the largest of the goose tribe which has fallen under our notice; it is of the size of the swan, and it often weighs more than twenty-five pounds. We have now in our possession one pair which we purchased for a gentleman in South Carolina, which will weigh, in common ordinary condition, over twenty pounds each. We once owned a gander that weighed twenty-four pounds. They are a noble bird, quite ornamental about the premises, and add much to the scenery, particularly if a sheet of water be near. When floating on its surface they have a stately, majestic appearance, and in their movements they much resemble the swan. They have a low, hollow, coarse voice, unlike that of any other variety.

The following description of this bird we take from "Buffon's Natural History."

"The appellation of Swan Goose, given by Willoughby to this large and beautiful bird, is very apt; but the Canada goose, which is at least as beautiful, has an equal right to the name; and besides, all compounded epithets ought to be banished from natural history. The Guinea goose exceeds all other geese in stature; its plumage is a brown-grey on the back, and light-grey on the fore side of the body, the whole equally clouded with rusty-grey, and with a brown cast on the head and above the neck; it resembles, therefore, the wild goose in its colors; but its magnitude and the prominent tubercle at the root of its bill, mark a small affinity to the swan; yet it differs from both by its inflated throat, which hangs down like a pouch or little dewlap; a very evident character, which has procured to these birds the denomination Jabotieres (from Jabot, the Crane). Africa, and perhaps the other southern countries of the old continent, seem to be their native abode; and though Linnæus has termed them Siberian geese, they are not indigenous in Siberia, but have been carried thither and multiplied in a state of domestication, as in Sweden and Germany. Frisch relates that, having repeatedly shown to Russians geese of this kind, which were reared in his court-yard, they all, without hesitation, called them Guinea geese, and not Russian or Siberian geese. Yet has the inaccurate denomination of Linnæus misled Brisson, who describes this goose under its true name of Guinea goose, and again, a second time, under that of Muscovy goose, without perceiving that his two descriptions refer precisely to the same bird."

"It is somewhat larger," says Brisson, "than the tame goose; the head and the top of the neck are brown, deeper on the upper side that on the under; . . . . . on the origin of the bill there rises a round and fleshy tubercle; . . . . under the throat also there hangs a sort of fleshy membrane." Add that Klien regards this goose of Muscovy or Russia as a variety of the Siberian, which, we have seen, is the same with a Guinea goose. "I saw," says he, "a variety of the Siberian goose, its throat larger, its bill and legs black, with a black depressed tubercle." Not only does this goose, though a native of the hot countries, multiply when domesticated in the coldest climates, it also contracts an affinity with the common species; and the hybrids which are thus bred take the red bill and legs of our goose, but retain of their foreign parents the head, the neck, and the strong hollow, yet loud voice. The clangor of these large geese is still more noisy than that of the ordinary kind, and they have many characters in common; the same vigilance seems natural to them. "Nothing," says Frisch, "can stir in the house during the night, but the Guinea goose will sound the alarm; and in the daytime they give the same screams if any person or animal enters the court; and often will pursue, pecking the legs." The bill, according to the remark of this naturalist, is armed at the edges with small indentings, and the tongue is beset with sharp papillæ; the bill is black, and the tubercle which rises upon it is vermillion. This bird carries its head high as it walks; and its fine carriage and its great bulk give it a noble air. According to Frisch, the skin of the little dewlap or pouch under the throat is neither soft nor flexible, but firm and hard. This account, however, scarcely agrees with the use which Kolbin tells us the sailors and soldiers at the Cape make of it. ("The wild geese at the Cape have been called Crop geese (oies Jabotieres). The soldiers and the common people of the colonies use their crops for tobacco pouches; they will hold about two pounds."—*Kolbin.*)

Fig. 59.

POLAND GOOSE.

There is a goose not uncommon in this country much resembling the Guinea goose, but not as large, known as the Poland goose; probably a cross of the Guinea goose and the Chinese goose next described.

CHINESE GOOSE.

The Chinese goose is not only brought from China, but from Guinea, the Cape of Good Hope in Africa, and Siberia, and it is also to be found in the Sandwich Islands, in the Pacific Ocean. It is some-

Fig. 60.

CHINESE GOOSE.

times entirely white. The variety from Guinea is known by its erect gait, and screaming, and is plentiful in this country.

This beautiful bird in its shape and motions in the water much resembles the swan. It again resembles this bird in other respects. She glides through the watery element with her neck beautifully arched, her head drawn in, her breast just settled in the water, her tail a little raised, giving a light airy appearance, moving on the water with apparently little or no exertion, and we may say in the language of the poet, "in all her actions dignity and grace." Her note is loud and shrill, and she utters it often when a stranger or an enemy appears. She is still more watchful than the African or Guinea goose. Nothing can stir about the premises in the night but she sounds the alarm. This, instead of the common, must have been the goose that is said to have saved the Capitol at Rome.

The plumage is grey on the back, and darker on the back side of the neck; front and under side of neck lighter and tinged with a fawn-color; wings and tail feathers dark, and under side of body light-grey. Feet, legs, and bill dark slate-color. She resembles the swan again in having a knob or fleshy tubercle on the base of the bill, at which also a small narrow white strip encircles the mouth.

Although a native of a warm climate, this bird appears very well naturalized in this country; the only or the greatest objection to them is their early laying, which often occurs in the dead of winter. Its beauty would merit a place in the poultry-yard. It will couple and breed with the common goose, but there would be no improvement on either side.

The Chinese goose is in size smaller than the common goose, and what they lack in size they make up in prolificness. "They are valued in this country," says Main, "as they are in their own, for their early breeding and aptitude to fatten. They begin laying at the end of November, if the season be mild, and in January goslings are hatched; and if kept in a dry warm room may be fit for the table in April or May. They are, however, excessively noisy birds; their flesh less bright, and considerably less delicate than our common geese; they are also much smaller."

The specimen from which our portrait was taken, has been in our possession for several years. She was imported from China, and we obtained her direct from the ship, and is the one before spoken of in a previous page.

Several gentlemen on Long Island, we have been informed, have attempted, and in some cases have succeeded in domesticating the *Brant*, which in its wild state is highly estemed for the exquisite delicacy of its flesh. Domestication, however, does not appear to have improved it much, and its small size will scarcely render it, except for curiosity, an object of much attention, particularly in the vicinity of their haunts.

HABITATION.

In selecting a situation for a goose-house or pen, all damp must be avoided; for geese, however much they may like to swim in water, are fond at all times of a clean dry place to sleep in.

It is not good to keep geese with other poultry; for when confined in the poultry-yard they become very quarrelsome, harass and injure the other fowls; therefore it is best to erect low sheds, with nests partitioned off, of suitable size to accommodate them;

and there should never be over eight under one roof; the large ones generally beat the smaller, in which case they should of course be separated, one from the other, by partitions extending out some distance from the nests.

The nests for hatching should be made of fine straw, of a circular shape, and so arranged that the eggs cannot fall out when the goose turns them. From fifteen to twenty eggs will be as many as a large goose can conveniently cover. The ganders remain near when sitting, and seem to watch them as a kind of sentinel, and wo be to man or beast that dares approach them; and they seem very anxious to see the young ones, that are to be born, make their appearance.

Incubation lasts from twenty-eight to thirty days, and not two months, as some state, and the hen should have water placed near her, and be well fed as soon as she comes off the nest, that she may not be so long absent as to allow the eggs to cool, which might cause her to abandon her task.

After twenty-eight or twenty-nine days' incubation, the goslings begin, but frequently at an interval of from twenty-four to forty-eight hours, to chip the shell.

Like turkey-chickens, goslings must be taken from under the mother, lest, if feeling the young ones under her, she might perhaps leave the rest of the tardy brood still unhatched. After having separated them from her, they must be kept in a basket, lined with wool and covered with cloth; and when the whole of the eggs are hatched, may be returned to the mother. The male seems to evince the same solicitude for the young as the mother, and will lead and take equal care of them. We once had a gander of the Chinese variety that actually took a brood of goslings from under a common goose, and brought them up with equal care.

On the second day after they are hatched they may be let out after the dew is off, if the weather is warm, but care must be taken not to expose them to the scorching rays of the sun, which might kill them. All authors seem to agree on the proper food to be given them, which is coarse barley meal, bruised oats, bran, crumbs of bread soaked in milk or curdled milk, lettuce leaves chopped fine, or crusts of bread boiled in milk. In this country, Indian meal moistened with water, is generally given, but in our experience we have found it too laxative, and to counteract the effect we have moistened it with boiled milk, and occasionally added chives chopped fine. It is our opinion, however, that more goslings are killed by over-feeding than by starving. A person who is curious in these affairs, informed us that he had been most successful when he let the goslings shift for themselves, if the pasture was good. We tried a brood that way and succeeded well. Grass seems to be their natural food, and by following nature in all cases, with animals, and more especially with fowls, we have generally succeeded best.

After they are three or four weeks old they may be turned out in a field or lane containing water. If their range is extensive they must be looked after, as the goose is apt to drag the goslings until they become cramped or tired, some of them squatting down and remaining at evening, and are seen no more.

After the goslings are pretty well feathered they are too large to be brooded under the mother's wings, and will sleep in groups by her side, and must be supplied with good and renewed straw to sit on, which will be converted into excellent manure. Being now able to frequent the pond and range the common at large, the young geese will obtain their own living; and if favorably situated, nothing more need be allowed them excepting the vegetable produce of the garden. We have however found it a good practice to feed a moderate quantity of solid food to the young and store geese, by which means they are kept in a growing and fleshy state, and attain a larger size; the young ones are also forward and valuable for breeding stock. Besides, feeding them, especially in the evening, on their return, attaches them to their home.

There is one thing the author has learned by sad experience, and that is, it will not answer to confine goslings in a small yard; they need exercise and a pasture to range in. We had a fine brood of fourteen, nearly feathered, confined in our poultry-yard with other fowls. We occasionally found one sitting by or on the water stupified, dumpish, and no inclination to eat or stir, and would remain so for one or two days and then die. After losing three in this way, we turned them out and let them range over the pasture and visit the pond, and never lost one afterwards.

DISEASES.

"Prevention is better than cure," so says the proverb. Cold and fogs are extremely against geese; therefore, when young, care should be taken not to let them out but in fair weather, when they can go to their food without a leader.

They are particularly subject to two diseases; the first a looseness or diarrhœa, for which Main recommends hot wine in which the parings of quinces, acorns, or juniper berries are boiled. The second is like a giddiness which makes them turn round for some time; they then fall down and die, if they are not relieved in time. The remedy recommended by Main, is to bleed the bird with a pin or needle, by piercing a rather prominent vein situated under the skin which separates the claws.

Another scourge to goslings, are little insects which get into their ears and nostrils, which fatigue and exhaust them; they then walk with their wings hanging down, and shaking their head. The relief proposed is to give them, on their return from the fields, some corn at the bottom of a vessel full of clear water; in order to eat it, they are obliged to plunge their head in the water, which compels the insects to fly and leave their prey.

PLUCKING.

Old, or what are termed stock-geese, may be plucked three, and in some seasons four times, allowing six weeks interval, without inconvenience. Many are of opinion that it was directly injuring the health of geese to pluck them. This operation, however, if done in a dexterous manner, and taking place before the moulting season, a disease common to all birds, is followed by no inconvenience. One crop of feathers may be taken from the goslings, and some think it an advantage to them, but that could hardly be expected; and it should be deferred till the goslings are three months old, before they are subjected to this operation, especially to those intended to be killed early, as they would get lean and lose some of their good qualities. Precaution should be taken when the goslings are just plucked, not to suffer them to go into the water, but merely give them drink for one or two days till the skin is closed. Food has a great influence over the quality of the down and feathers, as also the care that is taken of the geese. Great precaution is necessary; the feathers always bring away with them a kind of fat, which would give them a disagreeable smell and perhaps spoil, if this was not prevented by putting them in the oven after the bread is taken out, and keeping them in a dry airy place. One pound of feathers is generally estimated to be the produce of a common goose; the Bremen and African will give more and of a superior quality. Lean geese yield more down than fat ones. None but feathers taken from live geese, or those just killed, should be carried to market; in the last instance they must be picked before the bird is entirely cold; the feathers are infinitely better for it.

On the subject of plucking the living geese we would willingly be silent; the torture experienced by the poor fowl from the too frequent unskilfulness and want of dexterity of the operator. The skin and flesh are sometimes so torn as to occasion the death of the victim; and even when the geese are plucked in the most careful manner they lose their flesh and appetite; their eyes become dull, their wings heavy and drag on the ground, and they languish in a most pitible state, during a longer or shorter period. Great mortality often occurs in flocks of geese, from sudden and imprudent exposure to cold, after being stripped, and more especially during severe storms and sudden atmospheric vicissitudes.

A writer in one of the magazines remarks humanely on the cruelty of picking geese, and proposes the following remedy. "Feathers are of but a year's growth, and in the moulting season they spontaneously fall off, and are supplied by a fresh fleece. When, therefore, the geese are in full feather, let the plumage be removed, close to the skin, by a sharp scissors, clipping them off as sheep are shorn; they will be renewed at moulting in the usual course of nature. The produce would not be much reduced in quantity, while the quality would be greatly im proved, and an indemnification be experienced in the consciousness of not having tortured the poor bird, and in the uninjured health of the fowl and the benefit obtained in the succeeding crop. After this operation shall have been performed, the down from the breast may be removed by the same means."

FATTENING.

"It is the same with the goose," says Main, "as with every other bird that is fattened up; that moment must be laid hold of, when come to a complete plumpness, they would soon get lean and die if they were not killed." Meal and skimmed milk will soon do the business: after ranging in the grain stubbles but little else will be required. These are called "*green geese*," and are most esteemed by the epicure; they will then be about six weeks old, tender and fine."

The writer of the article on poultry, in "*Baxter's Library of Agriculture*," recommends steamed potatoes, with four quarts of ground buckwheat or oats to the bushel, mashed up with the potatoes, and given warm. This, it is said, will render geese, cooped in a dark place, fat enough in three weeks.

The French method of fattening is detailed very copiously by M. Parmentier. "The whole process," says he, "consists in plucking the feathers from under the belly; in giving them abundance of food and drink, and in cooping them up more closely than is practised with common fowls; cleanliness and quiet being above all indispensable. The best time is in the month of November, or when the cold weather begins to set in. When there are but a few geese to fatten, they are put in a cask, in which holes have been bored, and through which they thrust their head to get their food; but as this bird is voracious, and as with it hunger is stronger than love of liberty, it is easily fattened, provided they are abundantly supplied with wherewithal to swallow."

The Romans considered the liver of the goose a great dainty, and to increase its size they fed them sixteen days on a paste of Turkey figs, stamped and beaten up with cream; their livers would thus be brought to table, each weighing three or four pounds. Equal parts of the meal of oats, rye and pease mixed with skimmed milk, form an excellent feeding article for geese and ducks.

The grand object of preparing, not geese only, but all kinds of poultry, for market in as short a time as possible, is effected solely by paying unremitting attention to their wants; in keeping them thoroughly clean, in supplying them with proper food (dry, soft and green), water, exercise, ground, &c. They should be fed three times a day: and it is truly astonishing how soon they acquire a knowledge of the time.

CHAPTER XI.

THE DUCK.

Nature and propensities—varieties—common—crested or top-knot—Muscovy—canvass-back—habits—duck's eggs—fattening—enemies of the duck.

THE Duck possesses many excellent qualities. They were great favorites with the ancients, from the mildness and simplicity of their character, from their great fecundity, laying a great number of eggs, from the cheapness and ease with which they are provided for. The feathers of the white sort are nearly as good as those of the goose.

Ducks thrive best, and are cleanest in the neighborhood of water, such as a pond or stream of water, as where there is an abundance of water they will find the greater part of their living. They are the most industrious of all the fowl tribe, and we have often gazed on them with admiration to see them sputter in shallow and dive down in deep water. Ducks are carnivorous as well as granivorous; they will thrive on flesh and garbage of any kind, like the chicken; yet water insects, weeds, vegetables, corn and peas, are its general food; they are also very fond of fish, and will greedily devour it even when part decomposed; this of course will impart a bad flavor to the flesh if continued.

"It is a mistaken notion," says Ames, "to suppose that ducks *must* have a pond or run of water; they will do very well where there is none. A small pan or shallow tub sunk in the ground and placed so as to receive the waste water from the pump or well, will afford every necessary arrangement." But from our experience we are satisfied that where there is no piece of water, they will *not* do as well, nor do they appear as beautiful.

The inoffensive and harmless character, the social and conversational qualities of ducks, render them not only pleasant but profitable animals to keep, and the contrast between them and chickens, in their nature and habits, is much in their favor. Of the kind and social nature of the duck, the following is related by Mowbray:

"We had drawn off for the table the whole of a lot of ducks, one excepted. This duck immediately joined a cock and hens, and became so attached to them, that it never willingly quitted their company, notwithstanding some harsh usage, particularly from the cock. It would neither feed nor rest without them, and showed its uneasiness at their occasional absence by continual clamor."

We have an individual duck of the crested variety, which, after losing its mate, would keep with a few particular fowls during the day, and at night when the fowls went to roost, she would follow up the stairs into the second floor, and sit as near the fowls as she could get. But after we had placed a few of the large African geese in the yard, she left the hens and contracted an intimacy with the geese, keeping constantly with them.

The manners and actions of the duck, whether upon land or water, are curious and pleasant to contemplate. Their regular afternoon parade and march in line, the elder drakes and ducks in front, from the pond homewards, is a beautiful country spectacle, to be enjoyed by those who have a relish for the charms of simple nature. A parcel of ducks, which had been accustomed to their liberty, were for some particular reasons shut up for several hours. On the door of their house being opened, they rushed out, threw themselves into rank and file, and marched with rather a quick step, three or four times round a certain space, constantly bowing their heads to the ground, then elevating them and fluttering their wings; the ceremony finished, they quickly adjourned to the water. We have laughed a thousand times at the conceit with which my boyish imagination was impressed, namely, that the act we had witnessed was nothing less than a duckish thanksgiving for their deliverance.

There are numerous varieties of the duck, of great diversity of size and color, though it is not usual to domesticate, except for curiosity, more than two or three of these. Some naturalists, it is said, comprehended upwards of one hundred species. Main, in his " Treatise on Poultry," gives thirty-one varieties, viz.:—The curve-billed duck; the membraneousbilled duck; the spotted-billed duck; brown duck; dusky duck; lobated duck; bushy-headed duck; harlequin duck; the red-crested duck; Domitian duck; Spanish duck; Georgia duck; the clucking duck; the soft-billed duck; Damietta duck; the little brown duck; summer duck; crested duck; the Iceland duck; long-tailed duck; Monacha duck; Muscovy duck; the Pira duck; the brown duck; the regal duck; the wild duck; the black-tailed whistling duck; the western or stellar duck; the cinnamon-colored headed duck; the grey-headed duck; and the Jamaica shoveller. In his catalogue, he has omitted the following: the Rhone duck; the white Aylesbury duck: and the canvass-back duck.

We will now make it our business to give the history of the domestic ducks, and a few of the wild varieties.

THE COMMON DUCK.

This duck is so well known that it hardly needs a description. It is generally dark brown or grey, and the wings and throat sometimes ornamented with changeable purple. The drakes of all sorts may be distinguished by the curled feathers in their tails; they are good layers (one the property of Mr. Morrel, laid an egg every day, for eighty-five successive days—*Mowbray*), good nurses and hardy. Mowbray also relates, that "a duck has been known to lay in the autumn, during forty-six nights successively, after which she continued to lay every other night."

Mowbray says, " The only variety of the common duck among us (England), is the Rhone duck, imported from France, generally of a dark-colored plumage, larger size, and supposed to improve our breed. They are of darker flesh, and more savory than the English ducks, but somewhat coarse."

" The white Aylesbury ducks are a beautiful and ornamental stock," continues Mowbray, " matching well in color with the Embden (Bremen) geese. They are said to be early breeders; vast quantities are fattened for the London markets, where they are in great demand ; many families in Bucks derive a comfortable living from breeding and rearing ducks, the greater part of which, the early ones, at all events, are actually brought up by hand. The interior of the cottages of those who follow this occupation presents a very curious appearance to the stranger, being furnished with boxes, pens, &c., arranged round the wall for the protection of the tender charge of the good wife, whose whole time and attention is taken up with this branch of domestic economy."

We once had a small flock of the white Aylesbury ducks, and were much pleased with them, and only abandoned them for the more fanciful variety of the

CRESTED, OR TOP-KNOT DUCK.

This is a beautiful and ornamental variety. We have had them pure white, black, and mixed with black and white, with large turbans or crests. The duck from which our portrait was taken, is a fair

Fig. 61.

CRESTED, OR TOP-KNOT DUCK.

specimen of the pied ones. The white are considered the most beautiful, as they have yellow beaks and legs. They are early breeders; lay well, and hatch well; but are too valuable at present to kill, as they bring in market from one to two dollars each. We are not advised of the origin of this duck. Main speaks of the "red-crested duck" from New Zealand, but which is not common there. A red crest grows on the head; a very glossy black grey is predominant on the back; and a deep greyish soot color on the belly; the bill and the legs are lead color, the irides golden.

Latham also speaks of the crested duck, and says, "This inhabitant of the extremity of America, is of the size of the wild duck, but is much longer, for it measures twenty-five inches in length; a tuft adorns its head; a straw-yellow, mixed with rusty-colored spots, is spread over the throat, and front of the neck; the wing speculum blue beneath, edged with white; the bill, wing, and tail are black; the irides red, and all the rest of the body ashy-grey."

Our summer duck, more commonly known by the name of "wood duck," received its name in consequence of its alighting on the branches, and building its nest in the hollows of trees; has a bunch of long, silky feathers, varied with white, bright green, and purple, growing on the head of the male, and forms a superb aigrette, which nods backwards and forwards on the head. This duck, which to beauty adds a fine-flavored flesh, when it feeds on grain, acorns, and beech-mast, is of a shy mistrustful disposition; nevertheless, when taken young, it can be domesticated, and our only surprise is that it has not been domesticated for its beauty, if nothing else.

Now, whether the domestic crested duck has been produced from a cross of either of the foregoing varieties, is a question we are unable to determine. If it sprung from either, its size would indicate that the one mentioned by Latham would appear the most likely to produce them.

Fig. 62.

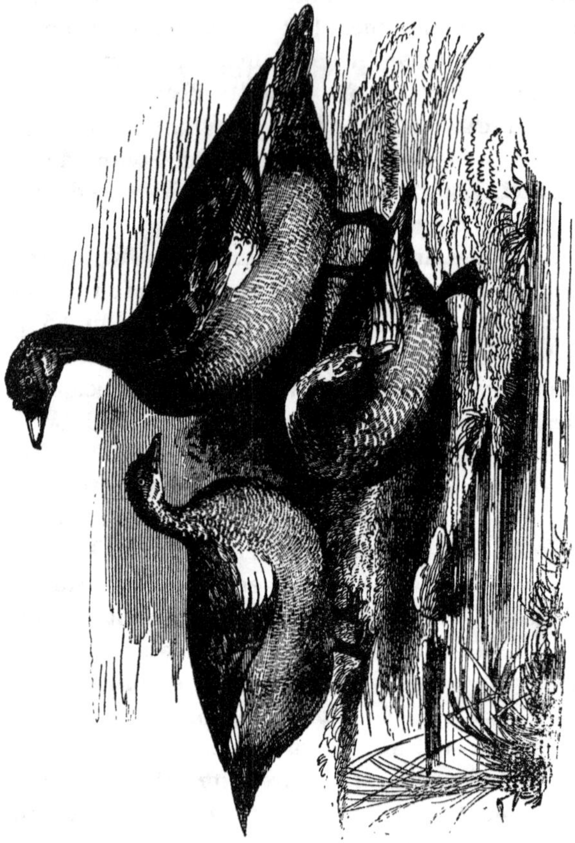

THE MUSCOVY DUCKS.

Our portraits were furnished by a friend from specimens in his own yard, who says in a note, accompanying them, " they are all of the same color, though I believe not necessarily so; a very dark rich blue-black, *prismatic*, with every color of which blue is a component, and a white bar is on the wing, some white about the head and neck, and I presume are to be found of all colors; the feathers, on the back of the male, are somewhat plume-like; the legs and feet dark," &c., &c.

It is asserted by authors, that the epithet musk has been given to this duck, because it exhales a strongish musky scent, owing to a humor which filtrates from glands placed near the rump. To take away this musky smell, the flesh of the rump must be taken off, and the head cut off, as soon as a bird of this species is killed. It is then a very good dish, and as succulent as the wild duck, and is occasionally met with on our tables.

The French naturalists assert that the Muscovy duck is a distinct species, and not a variety. It is much larger than the common duck, and is distinguished by a caruncled membrane of a red color, which Brun compares to a cherry, covering the cheeks, and extending behind the eyes, and swells at the root of the bill; this tubercle is wanting in the female, as also the tuft of narrow feathers, and rather twisted, which hangs behind the head of the male, which stands erect when excited. She is also smaller; both stand low on the legs, have short claws, and the inner ones crooked; are a clumsy bird on the ground, light on the wing, and will perch on fences, etc.

"In a wild state," says Brun, "the drake is of a brownish black color, with a broad white patch on the wings, the female being smaller and more obscurely colored." In the domestic state, it exhibits every variety of color, like a common duck. "At one time," says Brun, "the male is white, at another, the female white; in other instances, both male and female are black, and again of great diversities of color; but they are commonly black, variegated with other colors." The black are glossed with green on the back, and changeable, as they are exposed to the rays of the sun.

The Muscovy duck, it appears, is only found in a wild state in South America. Marcgrave has observed it in the Brazils; it is also a native of Guiana. Travellers assert that these birds perch on the large trees that border rivers and marshes, similar to terrestrial birds; they build their nests there, and as soon as the ducklings are hatched, the mother takes them one by one, and drops them into the water; laying takes place two or three times a-year, and each is from twelve to eighteen eggs, quite round, and of a greenish white; the moulting season begins in September, and is sometimes so complete, that the ducks, finding themselves almost entirely destitute of feathers, are unable to fly, and let themselves be taken alive by the Indians. These birds are as shy as our wild ducks, and it is by surprise alone that they are to be shot.—*Main.*

The Muscovy duck is easily fattened, and a prolific breeder, and hence, and though it is also a voracious feeder, it may be rendered profitable to rear. The male is very salacious, and pairs readily with the common duck, producing, by the cross, a hybrid or mongrel breed, which improves the size of the one and the delicacy of the other, but the mongrels, like that of the wild and tame goose, will not breed. The female will also, though not so readily, pair with the common drake. The hybrid has a deep green plumage, and is destitute of the red caruncled membrane on the cheeks, as well as the musky odor of the rump gland of the Muscovy duck. Out of one hundred eggs of this hybrid sort, M. Parmentier was able to succeed in hatching scarcely twenty ducklings; and hence to keep up the stock, Olivier de Serres advises to continue crossing every year, by keeping a sufficient number of Muscovy drakes with the common ducks.

Scaliger and Olivier de Serres have given out this duck was dumb.

The Muscovy drakes are often very cross and quarrelsome with other poultry, and we have known them to attack small children, particularly when they happen to have any food in their hands, and for which reason we have abandoned the rearing of them.

THE CANVASS-BACK DUCK.

This variety is a native of this country, and are found on the waters of the Potomac and Susquehannah rivers, on the banks of which they breed extensively. They are supposed also to breed on the borders of the Northern Lakes, or of Hudson's Bay; they frequent the waters of the Chesapeake, and are abundant on the Mississippi. Every one has heard, if they have not tasted, of the superior quality of their flesh. It is well ascertained that they feed on a bulbous root of a grass, which grows on the flats in the fresh water of these rivers, and has very much the color and flavor of garden celery; it is to this food that has been attributed, and, we believe correctly, the peculiarly delicious taste of their flesh. They feed in from five to ten feet of water; they are expert divers, and with great strength and agility, seize the grass near the bottom, bringing it up root and branch to the surface, where they bite off the root, leaving the long herbaceous part to float on the water.

This duck probably received its name from the peculiar color of the feathers on the back, which very much resembles coarse canvass. Their heads and necks are of a dark cinnamon or maroon color. There is a species of duck, which very much resembles the canvass-back, that are found in the Sound and Bays in the vicinity of New York, but they are shorter in body, and not as large; their bill is flatter and broader, and their flesh much inferior. The color of their plumage is much the same, but not so brilliant.

We are informed that attempts have been made to domesticate the canvass-back duck, but with what success we have not been advised. We have seen in the yard of Mr. Geo. Law, of Baltimore, a half-blood—a cross of the canvass-back and the common duck. She was shy, and seemed to retain many of the wild habits, and did not seem to care for the company of the other ducks.

HABITS OF THE DUCK.

We will now endeavor to point out to those who are desirous of making it their business to rear them, the most essential qualities of ducks, the means of deriving every advantage that can be obtained from them.

" The duck," says Main, " is undoubtedly a pretty good resource to the inhabitants of the country; it lives and multiplies in the midst of our habitations, requires little care, even in its first stage; provided it has at hand a river, a pond, a stream of water, a mud-hole, a slough, it is very little consequence which, moisture is its element; it could not answer anywhere but in cold aquatic places; it would be fruitless to persevere in the desire of bringing up ducks in dry and barren places; their flesh would neither be so tender nor so sweet; in this case it is better to take in preference to them some other birds, to whom the localities are better suited, to come into the views which are in contemplation."

The size of the duck is less than the drake; her plumage is neither so beautiful nor so bright; she makes much more noise; her voice is much louder and sharper; she is more loquacious than the male, whose voice is much weaker, monotonous, and resembles a whisper.

A single drake is enough for six or eight ducks; they begin to lay, if well fed in the winter, the latter part of March or first of April, if they are lodged in a dry comfortable place. They must then be closely looked after, for they are very careless, and deposit their eggs wherever they happen to be, in the water, in shady and secluded places, even after having concealed them from the vigilance of the person who has care of them; they hatch them secretly, and some fine morning bring their young brood to the house to ask for food, without requiring further trouble. It is prudent, when the spring is at hand, to give them food, three or four times, but little at a time, and always in places where it is wished they should lay, in placing their nests where they once have lain.

Where they are kept in any considerable quantities they should have accommodations of their own, and much on the plan of a goose-house, though it need not be so large. It should be secure at night from prowling animals, such as skunks, foxes, weasels, minks, and rats; the latter being the most destructive enemy we ever had to contend with. The walls and roof should be low and thickly thatched with straw, for warmth in winter, with the necessary openings for ventilation, cleaning, and egress and ingress of the ducks.

Ducks' eggs.—The common duck could lay from fifty to sixty eggs one after the other, daily, from the month of March till May, if the sitting did not intervene and interrupt the fowl. They are as nourishing though not so good-flavored as those of the common fowl, and answer equally as well for custards, &c.; are somewhat larger, and the shell seems smoother, and not quite so thick. Their color is usually greenish outwardly; there are some of them dull white; the yolk is large and pretty deep-colored.

Hatching.—Ducks are not generally inclined to sit; but to induce them to do so, towards the end of laying, two or three eggs are usually left in each nest, being careful every morning to take away the oldest in order that they may not spoil. From nine to eleven eggs are allowed her, according as she is able to cover them.

The only time the duck requires some care, is when she sits; as she has but little time to spare to procure her meals, food and water should be placed near her; and she is content with it, let the quality be what it may; it has ever been remarked, that when she was too well fed, she did not sit well; for that reason she should be portioned.

Incubation, like the goose, lasts thirty days; and the first broods are generally the best, because the warmth of summer helps much to bring them about;

the cold always prevents the late broods from getting strong, and giving as large ducks.

"The duck is reproached," says Main, "with letting her eggs get cold when she sits. Yet Reaumur asserts, he had a duck of the common species, which appeared still more uneasy about the cooling to which the eggs were going to be exposed while she was taking her food, than hens appeared to be for theirs; she only left the nest once a day, towards eight or nine in the morning; and before she left it, she covered them over with a layer of straw, which she drew from the body of the nest, to secure them from the impression of the air. This layer, above an inch thick, secured the eggs so well that it was quite impossible to guess that they were there."

To be sure, every duck of the same species is far from giving the same proofs of so much foresight, for the preservation of the warmth of her eggs, as the one above alluded to. It often happens that they let them cool. Besides, hardly are the ducklings born, when the mother takes them to the water, where they dabble and eat at first, and many of them perish if the weather is cold.

For the foregoing reasons, it is well to set hens on ducks' eggs; being more assiduous than ducks, these foster-mothers have more affection for their young, will hatch and guard them with more attention; and as they are unable to accompany them on the water, for which ducks show the greatest propensity, as soon as they are excluded, they follow the mother hen on dry land, and get a little hardy before they are allowed to take the water without any guide.

On hatching there is no necessity of taking away any of the brood, unless some accident should happen; and having hatched, let the duck retain her young upon the nest her own time. On her moving with the brood, prepare a coop and pen upon the short grass, if the weather be fine, or under shelter if otherwise; a wide and shallow dish of water, often to be renewed, near by them; their first food should be crumbs of bread moistened with milk; curds, or eggs boiled hard and chopped fine, is also much relished by, and is good for them. After a few days, Indian meal *boiled*, and rolled between the hands, and if boiled potatoes, and a few chives or lettice chopped fine be added, all the better. As soon as they have gained a little strength, a good deal of potherbs may be given them raw, chopped and mixed with a little bran soaked in water, barley and boiled potatoes beat up together. They are extremely fond of angle worms and bugs of all kinds, and for which reason they may be useful to have a run in the garden daily. All these equally agree with young ducks, which devour the different substances they meet with, and show, from their most tender age, a voracity which they always retain.

The period of their confinement to the pen, depends on the weather and strength of the ducklings. Two weeks seems the longest time necessary; and they may sometimes be permitted to enjoy the pond at the end of the week, but not for too great a length of time at once, least of all in cold weather, which will affect and cause them to scour and appear rough and draggled. In such case, they must be kept within a while, and have an allowance of stronger food. The straw beneath the duck should be often renewed, that the brood may have a dry and comfortable bed; and the mother should be plentifully fed with solid corn, without an ample allowance of which ducks are not to be reared or kept in perfection, although they gather so much abroad.

When ducks are hatched under hens they should not be allowed to go to the pond, until they become pretty hardy by remaining on the land; their natural propensity soon draws them to the water, and as soon as they see it they plunge into it, to the great alarm and distress of their foster-mother, who cannot follow them, as is sufficiently visible, and in fact, she is injured by the anxiety she suffers in witnessing the supposed perils of her children venturing upon the water.

A writer in the Southern Agriculturist, in speaking on the subject of rearing ducks, says, " These birds being aquatic in their habits, most persons suppose they ought to give the young ones a great deal of water. The consequence is, they soon take colds, become droopy, and die. This should be avoided. Ducks, when first hatched, are always inclined to fever, from their pinion wings coming out so soon. This acts upon them as teething does on children.

The young ducks should, consequently, be kept from everything which may have a tendency to create cold in them. To prevent this, therefore, I always give my young ducks as little water as possible. In fact, they should only have enough to allay their thirst, and should on no account be permitted to play in the water. If the person lives near the city, liver and lights should be procured, and these should be boiled, and chopped up fine, and given to the young ducks. Or, if fish, crabs, oysters, or clams, can be procured, these should be given. In case none of these can be obtained, all the victuals should be boiled before feeding. Boiled potatoes mixed with hominy are also excellent. Half of the ducks which are lost, are because raw food is given them. To sum up all in a word, if you wish to raise almost every duck that is hatched, give them little water, and feed them on no food which is not boiled. By observing this plan I raise for market, and for my own table, between two and three hundred ducks every year."

Ducks, when young, are exposed to many dangers and mishaps. Their waddling gait quite unfits them for running from a foe on land, and they are but too apt to be trodden on by horses, cattle, and even by the foot of man.

Care must be taken that the water where they are at liberty to go, contain no leeches, which occasions the loss of the ducklings, by sticking to their feet. We have also suffered some loss from the mud turtle, which infests some streams. We were once passing near a small stream, and hearing the cries of a gosling, we hurried to the bank, and found its feet apparently entangled; on grasping it, we found something hanging to the foot, and on raising it from the water, behold a snapping turtle had fastened to one of the legs, to which he adhered with the tenacity of a bull-dog. We threw him on the bank; he weighed ten pounds, and furnished us a repast, which would make an alderman's mouth water.

Boswell relates the following:—"To show that even in their congenial element, when skimming the surface of the water, under the watchful care of their mother, they are not free from danger, the following story is told by Waterton; "In 1815," says he, "I fully satisfied myself of the inordinate partiality of the carrion-crow for the young aquatic poultry. The duck had in her possession a brood of ten ducklings, which had been hatched about a fortnight. Unobserved by anybody, I put the old duck and her young ones into a pond, nearly three hundred yards from a high fir-tree, in which a carrion-crow had built its nest; it contained five young ones almost fledged. I took my station on the bridge, about one hundred yards from the tree. Nine times the parent crow flew to the pond, and brought back a duckling each time to its young. I saved a tenth victim by timely interference. When a young brood is attacked by an enemy, the old duck has nothing to defend it. In lieu of putting herself between it and danger, as the dunghill fowl would do, she opens her mouth and shoots obliquely through the water, beating it with her wings. During these useless movements, the invader seizes its prey with impunity."

Fattening.—According to Gervase Markham. pulse of any kind of grain will fatten ducks in a fortnight; we are not of that opinion, and we think if he had tried it, he would have found that his recipe was not always successful.

Ducks are fattened, either in confinement, with plenty of food and water, or full as well, restricted to a pond, with access to as much solid food as they will eat. They fatten speedily by mixing their hard meat, as an Englishman would say, with such variety abroad as is natural to them, more particularly if in good condition, and there is no check or impediment to thrift, from pining, for every mouthful tells and weighs. A dish of mixed food, if preferred to whole grain, may remain on the bank, or, rather in a shed, for the ducks. "I must here mention a fact," says Mowbray, "which I have either actually verified, or suppose that I have verified. Barley, in any form, should never be used in fattening aquatics, ducks or geese, since it renders their flesh loose, *woolly*, and insipid, and deprives it of that high savory flavor of brown meat which is its valuable distinction; in a word, rendering it *chickeny*, not unlike in flavor the flesh of ordinary and yellow-legged fowls." Oats and corn are the standard material for ducks and geese, to which may be added boiled potatoes and

Indian meal, or ship stuffs, mixed as it may be required. Liver boiled and chopped fine is a good condiment, and well relished by ducks. In England, they are fattened on ground malt, mixed up with water and milk.

When ducks are confined to fatten, or otherwise, it is well to give them sand, or brick pounded fine and mixed with their food, and occasionally earthworms. If their droppings are too loose and watery, mix a little forge water in their food; this will also cure the relax in any other sort of fowls.

A deceased friend of the author, who was very curious in these matters, and besides a lover of the good things of this world, used to feed his ducks, ten or twelve days previous to killing them, with celery chopped fine, to give them a flavor, which he assured us rendered their flesh but little inferior to the famous canvass-back ducks.

That the food on which fowls are fed has a tendency to impart a flavor to their flesh, and even to the eggs, is obvious from a fact related to us not long since, by a friend. He said some onions, partly decayed, were thrown into a yard where he had some fowls confined, of which they ate considerable. A few days after, he was surprised to find his eggs tasted so strongly of the onions that they could not be eaten. It is also well known that when fowls are fed on fish, their flesh has always a fishy taste.

Ducks are so very greedy, that they often devour a whole fish or a frog, which hurts them very much, if they do not immediately throw it up. They are particularly fond of meat, which they eat with avidity, even when tainted; slugs, spiders, toads, insects, all suit their ravenous appetite. They therefore are, of all the birds in the poultry-yard, those that do the greatest service in a garden, by destroying insects which do so much damage, did not their own voraciousness cause other inconveniences, which more than balance this advantage.

Cobbett advises feeding them—"grass, corn, cabbages, and lettuces, and especially buckwheat, cut when half ripe, and flung down in the haulm. This makes fine ducks. Ducks will feed on garbage and all sorts of filthy things; but their flesh is strong and bad in proportion. They are, on Long Island, fattened upon a coarse sort of crab, called a horse-foot, cast on the shores. When young, they should be fed upon barley-meal, or curds, and kept in a warm and dry place in the night time, and not let out early in the morning. It always does them harm; and if intended to be sold or killed young, they should never go near ponds, ditches, or streams. When you come to fat ducks, you must take care that they get no filth whatever. They will eat garbage of all sorts; they will suck down the most nauseous particles of all those substances which go for manure. A dead rat, three parts rotten, is a feast to them. For these reasons I should never eat any ducks, unless there were some mode of keeping them from this horrible food. I treat them precisely as I do my geese. I buy a troop when they are young, and put them in a pen, and feed them upon oats, cabbages, lettuces, and have the place kept very clean. My ducks are, in consequence of this, a great deal more fine and delicate than any others that I know anything of."

They live chiefly on grain scattered about the premises, the siftings and sweepings of barns, all sorts of mealy substances, the residue of breweries and boiling-houses, roots, fruits, everything, indeed, suits them, provided it be rather moist—in fact, nothing seems to come amiss to them.

Their weight, size, and flavor, depend much upon the manner in which they have been fed and fattened. The size of the duck varies much. There are some which, in the course of eight or nine weeks, reckoning from their birth, weigh as much as seven or eight pounds, while others, of the same age and species, do not come to half this weight. As this bird values its liberty very much, it is no less strange than true, that it fattens more readily and rapidly not only in confinement, but even when cooped up; repose and good living appearing to hasten even aldermanic obesity.

Enemies of the duck.—The most dreaded is the fox, to whose incursions the ducks are most liable, because they pretty commonly stray from home; and it cannot be hunted too much to get the country rid of it; and the ducks must be driven to the water in the morning, and brought back in the evening.

Minks and weasels will also destroy ducks; we had a very fine duck killed by a mink, just wounded in the neck, no other part of the body was touched. Skunks will also destroy them if an opportunity offers. But of all animals most to be dreaded for the ducklings is the rat. We have suffered more from their depredations than all other animals put together, and they are the hardest to be got rid of. To avoid them, the ducks as well as chickens, should not be cooped too near any building.

CHAPTER XII

INCUBATION.

Incubation—process—nature the best guide—artificial method—amusing anecdote—Reaumur's method of hatching by steam—Eccaleobion—Mowbray's plan—artificial mother—Reaumur's artificial mother—feeding and fattening poultry—feeding-houses

The process of incubation of the chicken is a subject curious and interesting to the student of nature. It generally takes twenty-one days to hatch a brood of chickens, although a close-sitting hen will sometimes hatch in eighteen days, if the weather is favorable. The expiration of the time should be carefully watched for; not that the chicken requires any assistance, but, on the contrary, interference with them is much more likely to injure than benefit them; a healthy chick will perform all that is required to free it from the shell. It is truly wonderful the power they possess while rolled up in so apparently a helpless mass; but so it is, and the head that makes the most exertion is placed so as to leave room for reaction, and to turn round, and thus to peck a circle as seen in Fig. 61, and breaks a circle around the large end of the shell, admitting the air by degrees, until it becomes gradually prepared to extricate itself

Fig. 63.

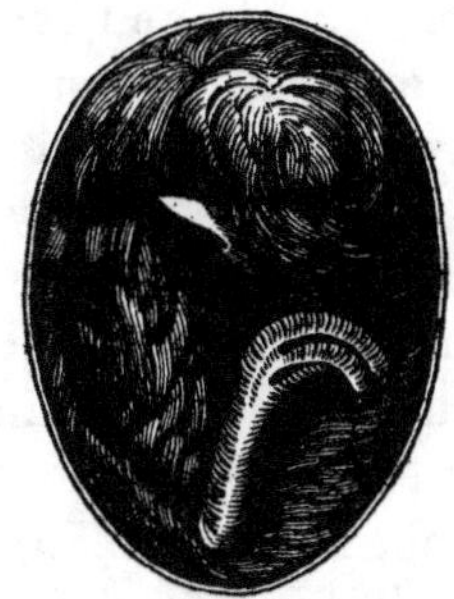

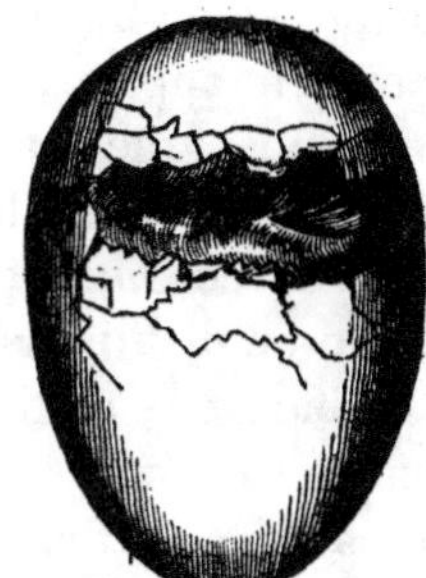

The following very interesting account of the wonderful changes which an egg undergoes in hatching, from the first day till its final exclusion, is particularly interesting, and is taken from an English Journal. By means of the Eccaleobion, and hatching ovens, many interesting facts have been discovered, and are described with great minuteness.

Fig. 64.

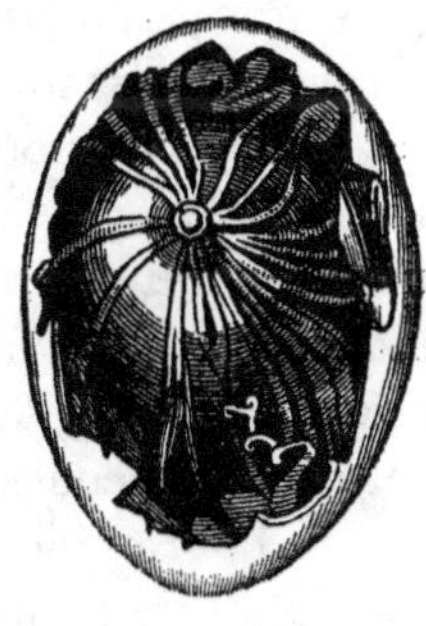

FIRST AND LAST STAGES OF THE CHICK.

"The hen has scarcely sat on her eggs twelve hours, before some lineaments of the head and body of the chicken appear. The heart may be seen to beat at the end of the second day; it has at that time somewhat the form of a horse-shoe, but no blood yet appears. At the end of two days, two vesicles of blood are to be distinguished, the pulsation of which is very visible; one of these is the left ventricle, and the other the root of the great artery. At the fiftieth hour, one auricle of the heart appears, resembling a noose folded down upon itself. The beating of the heart is first observed in the auricle, and afterwards in the ventricle. At the end of seventy hours the wings are distinguishable; and on the head, two bubbles are seen for the brain, one for the bill, and two for the fore and hind part of the head. Towards the end of the fourth day, the two auricles, already visible, draw nearer to the heart than before. The liver appears towards the fifth day. At the end of a hundred and thirty-one hours, the first voluntary motion is observed. At the end of seven hours more, the lungs and stomach become visible, and four hours afterwards, the intestines, the loins, and the upper jaw. At the hundred and forty-fourth hour, two ventricles are visible, and two drops of blood instead of the single one which was seen before. The seventh day, the brain begins to have some consistency. At the hundred and nineteenth hour of incubation, the bill opens, and the flesh appears in the breast; in four hours more the breast-bone is seen;

in six hours after this the ribs appear, forming from the back, and the bill is very visible, as the gall bladder. The bill becomes green at the end of two hundred and thirty-six hours; and if the chicken be taken out of its covering, it evidently moves itself. The feathers begin to shoot out towards the two hundred and fortieth hour, and the skull becomes gristly. At the two hundred and sixty-fourth hour, the eyes appear. At the two hundred and eighty-eighth the ribs are perfect. At the three hundred and thirty-first, the spleen draws near the stomach, and the lungs to the chest. At the end of three hundred and fifty-five hours, the bill frequently opens and shuts; and at the end of the eighteenth day, the first cry of the chicken is heard. It afterwards gets more strength and grows continually, till at length it is enabled to set itself free from its confinement.

"In the whole of this process we must remark that every part appears at its proper time; if, for example, the liver is formed on the fifth day, it is founded on the preceding situation of the chicken, and on the changes that were to follow. No part of the body could possibly appear either sooner or later without the whole embryo suffering; and each of the limbs becomes visible at the first moment. This ordination, so wise and so invariable, is manifestly the work of a Supreme Being; but we must still more sensibly acknowledge His creative powers, when we consider the manner in which the chicken is formed out of the parts which compose the egg.

How astonishing it must appear to an observing mind, that in this substance there should at all be the vital principle of an animated being! that all the parts of an animal's body should be concealed in it; and require nothing but heat to unfold and quicken them! that the whole formation of the chicken should be so constant and regular that exactly at the same time, the same changes will take place in the generality of eggs! that the chicken, the moment it is hatched, is heavier than the egg was before! But even these are not all the wonders in the formation of the bird from the egg (for this instance will serve to illustrate the whole of the feathered tribe); there are others altogether hidden from our observations, and of which, from our very limited faculties, we must ever remain ignorant."

In hatching of poultry, as in most other things, nature is the best guide. A hen is generally ill to please in the choice of her nest, and should always have a variety among which to choose. Generally several will lay in the same nest, and frequently two or more at the same time, and generally prefer the nest that contains the most eggs, and will always give the preference where at least there is one; hence the necessity of always leaving one in the nest as a nest egg. The hen and duck, if left to themselves, find some dry, warm, sandy hedge or bank, in which to deposit their eggs, forming their nests of moss, leaves or dry grass. In this way the warmth is retained in the nest for the few moments she devotes to her hurried and scanty meal. In imitating nature, in a degree, or approaching to that end, we would advise the adoption of ever-green boughs, as seen in page 99. It is a good mode to put in the bottom of the nests a sufficient quantity of dry sand and grass, moss, or cut straw. Wood ashes and tobacco stems have been recommended by some, as they produce the effect of destroying or preventing vermin, by which they are apt to be infested at that time.

It is well known that when a certain number of eggs has been laid, the hen shows an inclination to sit. She has a peculiar sort of clucking, and a feverish state ensues in which the natural heat of the hen's body is very much increased. The inclination soon becomes a strong ungovernable passion; which appears a blind instinct, for she will sit upon one egg or twenty, upon a piece of chalk or a stone; and if fresh eggs are supplied, she will sit for six weeks. In this state she flutters about, hangs her wings, bristles up her feathers, searches everywhere for eggs to sit upon; and if she finds any she immediately seats herself upon them.

These signs of an inclination to sit are not, however, sufficient to cause her to be entrusted with the duty of hatching, since all hens, particularly young ones, are not to be depended on. It is, therefore, a good practice to make trial of a hen by leaving her to sit on a nest for a few days, upon a few chalk-eggs; and if she continues to sit with constancy, the proper number of eggs may be placed under her.

During the period of incubation, a good sitter will not leave her nest for more than a few minutes at a time, to provide her food, and at intervals of from one to three days. So powerful, too, is this instinct, that they have been known to remain on their nests until they have perished with hunger. To prevent such an occurrence, it has been recommended to feed them daily in this situation; but from our experience it seems the best plan to let them follow the dictates of their own instinct, and when they desire food and water let them seek it in the poultry-yard.

"At this season, too, her whole nature is changed. From being peaceful and cowardly, she becomes a noisy termagant, fighting with all her female friends, and avoiding chanticleer as her most dreadful foe. Her notes change to a peculiar cluck, which lasts until her young brood can shift for themselves."

The number of eggs to be put under a hen must vary according to the size of the hen and the temperature of the weather. It is the common practice to set them upon odd numbers, as 11, 13 and 15, so that they will form a circle around the centre; but sometimes the hen may lay more eggs, or others lay in the same nest; we have, therefore, found it necessary to mark the eggs with ink, and if fresh ones should be laid, they could readily be distinguished and removed, as they would be too late in hatching.

For hatching, and to have the eggs productive, they must be fresh, and must not be exposed to noxious effluvia or moisture. Those intended for incubation should always be gathered with more care than if they were merely to be employed for aliment. They should be of an average size and ordinary form, avoiding very small eggs, which have generally no yolk, and those which are ill-shaped, or of equal thickness at both ends, as the latter are the usual shape of such eggs as have double yolks, which, though good for culinary purposes, are not so for hatching; for if they prove productive, the produce are generally monsters with two heads, four legs and the like. Instances have occurred, but rarely, when two and even three chickens were hatched from the same egg.

"If you wish for dark-colored chickens, you have only to select those eggs which have a light reddish-brown hue." Sketcherly observes, "I have generally found that the round egg produced the female, and those of oblong form the male; but this is no new discovery, for it was known more than two thousand years ago; for Mascall says, out of Columella, "If ye will have them to be females, take the roundest wrinkled egges, and also to have them all males, take the long rough egges. Also some doe chuse the egge with the hollow crowne in the side of the toppe for the females, and the crowne in the toppe under the shell for the males." This does not seem to be very intelligible, but it may perhaps refer to what M. Parmentier considers to be a discovery. "Formerly," he says, "pointed eggs were chosen to produce cocks, and round ones to produce hens;

now they are known by more certain signs. Examine the eggs by holding them between the eye and the candle, and if the vacancy caused by the air-bag at the blunt end of the egg, appear to be a little on one side, it will produce a hen; if this vacancy be exactly in the centre, it will produce a cock." Now, this all appears well in theory, but from our experience we have not found it so. Last year we were very desirous of rearing a large proportion of females, and for that purpose we selected *all* round eggs, rejecting the long and pointed ones; the result was, there were about an equal number of both cocks and hens.

It has generally been found that hens which are the best layers are the worst sitters. Those which we have found best adapted for that purpose, have skort legs, a broad body, large wings well furnished with feathers, their nails and spurs not too long or sharp.

After twenty-one days, a good sitter will bring out her chickens, and as soon as she becomes a mother her whole character is changed. All her former feelings and habits become absorbed in increasing maternal solicitude. She turns out to be frugal, generous, sober, reserved, courageous, and intrepid. She assumes, indeed, all the qualities of the cock, and even carries them to a higher degree of perfection. When we see her come into the poultry-yard, surrounded by her little ones, for the first time, it seems as if she was proud of her new dignity, and takes a great pleasure in performing her duty. Her eyes are lively, animated and constantly on the alert; her looks are so quick and rapid that she could take in every object at one glance; and she appears to discover at once the smallest seed on the ground, which she points out to her young ones; and in the air, if she discovers the bird of prey, she dreads for their sake, and giving them warning by a peculiar cry, she induces them immediately to hide themselves.

Incessantly taken up with their welfare, she excites them to follow her and to eat; she picks them food; she scratches the ground in search of worms which she gives up to them; she stops now and then, squats down, opens her wings and invites her tender brood to come and gather around and warm themselves beneath her. She continues to bestow these cares on them until they are quite feathered, when they are fit to shift for themselves.

The first day after hatching, the chickens do not want to eat, and should be left in the nest. The next day, the whole brood being hatched, the hen with them may be removed and placed in a box, with high sides, if the weather is cold or wet; or put under a coop, upon a dry sheltered spot, and, if possible, not within reach of another hen, since the chickens will mix, and the hens are apt to injure, and often kill such as do not belong to them. Nor should they be placed near other fowls, as they would rob them of their food.

Their first food may be eggs boiled hard and chopped fine, or curd broken fine; millet or coarse ground corn, commonly called "samp" at the North and "homminy" at the South, and fed sparingly, a little at a time and often at first, as from our experience, we are certain more chickens are destroyed by over-feeding than are lost by the want of it. We have remarked also that hens which stole their nests generally hatched all the eggs; and if suffered to seek the food for her chickens, if the season was somewhat advanced, she would, unless some casualty occurred, raise the whole brood, while with too much kindness or officiousness, not half would be raised. All watery food, such as soaken bread, or potatoes, should be avoided. If Indian meal is well boiled and fed not too moist, it will answer a very good purpose, particularly after they are eight or ten days old. Pure water must be placed near them, either in shallow dishes or bottle fountains as in page 123, that the chickens may drink without getting into the water, which, by wetting their feathers, benumbs and injures them. After having confined them for five or six days in the box, they may be allowed the range of the yard if the weather is fair. They should not be let out of their coops too early in the morning, or whilst the dew is on the ground, far less be suffered to range over the wet grass, which is a common and fatal cause of disease and death. Another cause of the utmost consequence to guard them against, is sudden unfavorable changes of the weather, more particularly if attended with rain. Nearly all the diseases of gallinaceous fowls arise from cold moisture.

At the end of four weeks the hen may be allowed to lead her little ones into the poultry-yard, where she will soon leave them and commence laying again.

It should be the aim to have some of the hens hatch as early as possible, so that the chickens will attain a good size by the first of July, and if fat, will return the best profit in market in proportion to their age and food consumed. They are naturally most fat at six weeks old or about the time they leave the hen, and have not run off their brooding flesh by exertion for food and by growth. Particular birds can be selected for breeding stock, as their color and form will be by that time apparent, so as to make the choice with safety; also, it will be easy to tell the males from the females.

If their keep costs nothing, and they are raised near or are convenient to a market, they may in some cases, be advantageously retained till the holidays, when they seldom fail to command a ready sale and a good price; but if a large number are raised, they will, of course, require to be marketed regularly. Of this, however, the farmer will be the best judge. In many cases it will be more advantageous to sell to the *dealers*, who travel the country in all directions with wagons prepared to take the fowls from the yard, pay cash price, sufficiently liberal to return a handsome profit to the breeder. These men also purchase feathers and eggs.

ARTIFICIAL HATCHING.

Fig. 65.

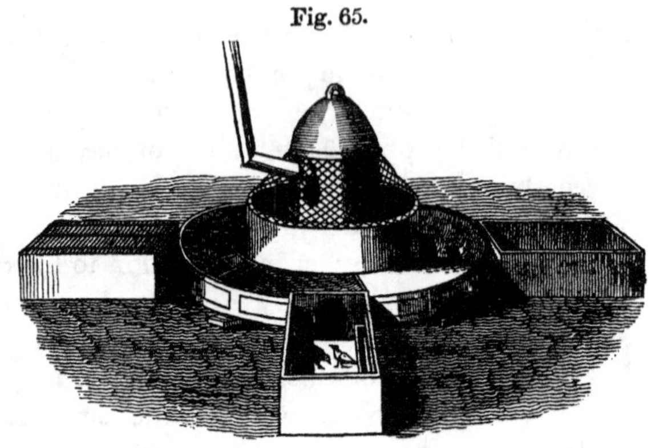

REAUMUR'S HATCHING APPARATUS.

The method of artificial hatching has never, we believe, until quite recently, been attempted in this country. It is, however, resorted to in some countries to a considerable extent, the demand for poultry being much greater than could be supplied by natural means. The first notices we have of hatching chickens artificially, without the aid of hens, are to be found in Aristotle and Pliny. The latter mentions that the Roman Empress Livia hatched an egg by carrying it about in her "warme bosome," and this probably gave origin to the "device of late to lay eggs in some warm place, and to make a gentle fire underneath of small straw or light chaff, to give a kind of moderate heat; but ever more the eggs must be turned by man or woman's hand both night and day, and so at the same time they looked for chickens and had them."

The art has been extensively practised in Egypt and China from an unknown period of time. In the former place immense numbers of eggs are hatched by heat in ovens or mammels; and officers are appointed by government to superintend the process, and receive a part of the produce as pay About the middle of January the ovens are inspected and repaired; and, as they are public, and as each has a circuit of fifteen or twenty villages, notice is given to the inhabitants, so as they may come and bring their eggs. As soon as a suitable quantity of eggs is collected together, they are put into the rooms that are to serve for the first brood; for the whole of the ovens are never employed at once on the same brood, but only one half of those which the building contains.

It is asserted by Barron that it is practised by the Chinese families who live constantly on the water. They deposit the eggs in sand, at the bottom of wooden boxes, placed on iron plates and kept moderately heated.

As there is no prospect of any of our countrymen entering into the business on the Egyptian method, we will not detain the reader by a description of these ovens; other and less expensive plans have been adopted. The same feat performed by Livia has been accomplished in New Jersey, where a lady

28

in Monmouth County patiently hatched two chickens, which she successfully raised. Some French ladies have in the same way proved themselves mothers of canaries and other birds.

"On this," says Ames, "I have heard an amusing anecdote, which I give as a hint for the advantage of those similarly situated. I am sure its veracity may be relied on. An industrious farmer's wife in New Jersey, had a husband or a thing she was obliged to call so, who was intemperate, hypochondriac and lazy; and after a debauch would sometimes remain in bed for several weeks, from which no persuasion or art could rouse him. His active *rib* hit upon an expedient to turn this to account, and immediately put it in practice. She procured a quantity of fresh eggs, and rolling them in wool and flannel, placed them around him in the bed so as to receive the necessary warmth, and in due time were brought forth a pretty flock of chickens. It was then further surmised, that finding him more useful in this capacity of an old hen than in any other, she encouraged him by '*tiny drops*,' to lengthen his periods of incubation."

But to return to the hatching apparatus. The French have made many experiments, among whom was the celebrated naturalist and French philosopher, M. Reaumur, who made a number under the immediate attention of the French king, and published the results in a treatise of five hundred pages, with plates. It states that he found the proper degree of heat to be about ninety degrees of Fahrenheit. He thinks it perfectly practicable.

Olivier de Serres, the father of French agriculture, describes a little portable oven, of iron or copper, in which eggs were arranged and surrounded with feathers, and covered with soft cushions, heat having been communicated by means of four lamps, but he says it was more curious than useful.

M. Dubois made an improvement upon M. Reaumur's suggestion of the stove, which is both easy and not expensive. "Below a chamber, ten feet by ten, with a low ceiling, and a door covered with old tapestry, M. Dubois places a metal stove with a pipe rising perpendicularly into the chamber to heat it. He burns balls of clay kneaded up with small coal, a sort of fuel common on the continent, and two pounds of coal added every five or six hours was found sufficient to keep up the proper temperature, which is ascertained by several thermometers. The eggs are placed in ozier baskets, suspended from hooks in the ceiling, each basket being dated on the day it is hung up. At the end of four or five days (but this is rather too soon) the bad eggs are removed. Having found that, towards the twelfth or fifteenth day of incubation, it is better to diminish the heat, M. Dubois lengthens the cords which suspend the baskets, so as to bring them nearer the floor of the chamber, where it is not so hot. He also moves the eggs daily so as to regulate their heat. It is not said whether this method has been much tried.

Instead of the dry heat of a stove or oven, M. Copineau makes use of hot water carried in a pipe along the floor of a chamber constructed for hatching. He also has pipes or flues for the purpose of ventilation and regulating the heat; while he places vessels of water in the chamber to render the air equally moist with that under a sitting hen.

Fig. 66

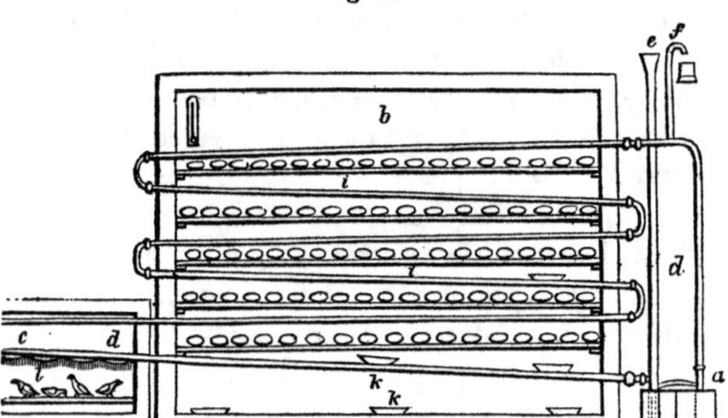

" The incubation of chickens by hot water is said to be the invention of M. Bonnemain of Paris. The above is a section of his apparatus, consisting of a boiler (*a*); a box or building (*b*) for hatching eggs; a cage or coop (*c*) for rearing the chickens; tubes (*d*) for circulating the hot water; a supply tube (*h*) and funnel (*e*), and safety tube (*f*). Supposing the water heated in the boiler, it will rise by its specific levity through the tube (*d*), move backwards and forwards through all the tubes, and return again to the boiler at (*k*), which is inserted in the top like the other, but passes down to its lower part (*l*). This circulating movement once commenced, continues so long as the water is heated in the boiler, because the temperature is never equal throughout all parts of the apparatus. We may readily conceive that a perfect equality of temperature can never exist, on account of the continual loss of heat, which escapes from the exertions of all the tubes. Meanwhile the temperature of the air enclosed in the box differs but little from that of the numerous tubes which traverse it; and as the bends of the tubes on the outside of the box afford but little surface to be cooled by the surrounding air, so the force of the circulation, which is always in the ratio of the difference between the temperature of the water's passing out of the calorifère and re-entering it, does not become greatly diminished, even after having expended a large portion of its heat on the outside of the box, in maintaining a gentle heat in the cage (*c*) adjoining to it. We see, therefore, the more the water is cooled which passes through the last circumvolutions of the tubes, the more active is the circulation in all parts, and consequently the more equal is the temperature of all the tubes which heat the box, and of the air within it; indeed, to prevent the loss of heat as much as possible, the boiler, and all those parts of the tubes which are placed on the exterior of the box, are enveloped in lists of woollen cloth. M. Bonnemain having thus applied these principles with so much skill, is always enabled to maintain in these boxes an equal temperature, varying scarcely so much as half a degree of Reaumur's thermometer ; but as if it was not sufficient to have thus far resolved the problem, he contrived that this degree of temperature in all parts of the stove should be maintained at that point which was found most favorable for promoting incubation. It was by means of an apparatus for regulating the fire that he attained this desirable object. The action of this regulator is founded on the unequal dilatation of different metals by heat. A movement is communicated near to the axis of a balanced lever, which lever transmits it by an iron wire to a register in the ash-pit door of the furnace. Combustion is by these means abated or increased."

"*When we would hatch chickens by hot water*, we light the fire and raise the temperature till we obtain that degree of heat in the box which is fitted for incubation; we then place the eggs near to each other upon the shelves with boilers to them. It is convenient not to cover, on the first day, more than a twentieth part of the superficies of the shelves, and to add every day, for twenty days, an equal quantity of eggs; so that we may obtain every day nearly the same number of chickens ; but which, nevertheless, may be occasionally regulated by the particular season of the year."

"During the first days of incubation, whether natural or artificial, the small portion of water contained within the substance of the egg evaporates through the pores in its shell; this is replaced by a small quantity of air, which is necessary to support the respiration of the chick; but as the atmospheric air which surrounds the eggs in the box at that degree of temperature is either completely dry or but little humid, so the chick would greatly suffer, or finally perish, from this kind of desiccation. The aqueous vapor which exhales from the breathing of the old fowls while hatching, in some degree prevents this ill-effect; but nevertheless, in dry seasons, the vapor is hardly sufficient, and thus, in order that the eggs may be better hatched in the dry seasons, the hens cover them with the earth of the floor of the granary. In artificial incubation, to keep the air in the stove constantly humid, they place in it flat vessels, such as shallow dishes (*i i*) filled with water. When the chickens are hatched, they are removed from the stove and carried to the cage (*c*), where they are fed with millet, and nestle under a sheep skin with wool on it (*j*) suspended over them. They also separate by means of partitions in the cage, the chickens as they are hatched each day, in order to modify their nourishment agreeably to their age. Artificial incubation is exceedingly useful in furnishing young fowls at those seasons when the hens will not sit, and in some situations, to produce, or as we may say indeed, to manufacture a great number of fowls in a small space."

A method somewhat similar to M. Bonnemain's,

to which a long Greek name has been given, has been put in operation at Pall Mall, London, and exhibited at 25 cents each person. In Chambers' Edinburgh Journal is a description of the "*Eccaleobion*." It is a room, one side of which is a large oblong case placed against the wall, divided into eight parts, each one of which is warmed by steam pipes, and which are used for hatching the eggs. The bottom of these boxes or parts, and indeed the whole, is lined with cloth, and is covered with eggs lying at a little distance from each other. There is a jug of water in each part to preserve a proper degree of moisture to the air in the divisions. The meaning of having eight boxes, is to ensure a batch of chickens every two or three days. Each part holds some two or three hundred eggs, or about two thousand in the whole. From twenty-one to twenty-three days are required to hatch the eggs, and as those are purchased in the market, from one-third to one-half prove worthless. None but new eggs should be used for the Eccaleobion.

When the chickens appear they are not immediately removed from the oven, but remain a few hours until dry, when they are taken from the oven and put into a glass case or box made shallow and the sash-lid easily removable. They are not fed for twenty-four hours after hatching, and the material then used is a coarse meal grit, which they pick up with great eagerness, instinct in this case supplying the want of the mother. They are kept in this case two or three days, when they are put into divisions on another part of the floor of the same large and warm apartment. At dusk they are put into a coop or box with a flannel curtain and covering, where they rest with as much quietness as under the wing of the mother. In the morning they are turned into the yard, which is cleaned and strewed with sand. When three weeks or a month old they fetch in market one shilling each. It thus appears that all that is necessary to form a chicken establishment is suitable rooms and a steady supply of the proper heat, fresh eggs and constant attention.

At the meeting of the Royal Agricultural Society at Bristol, in 1842, a small machine for hatching chickens artificially, was exhibited by the inventor, Mr. C. Appleyard of London. As the notice was unaccompanied by a description, we can only say such a machine was exhibited.

Mowbray, in his work on poultry, gives an account of a clumsy method which he tried for hatching chickens. He wrapped a number of eggs in wool, put them in a market-basket covered with flannel, and suspended this over a chafing-dish of charcoal in a chimney where there was no other fire, the chimney-screen being constantly kept fast to concentrate the heat. The degree of heat was judged of, every three or four hours, by feeling, and the eggs constantly turned and transferred from the centre to the circumference of the basket. About thirty or forty healthy chickens were, on a second trial, obtained from forty-five eggs, the first trial having been unsuccessful.

"Here," says Mowbray, "commenced the grand difficulty. The nurse chickens soon became weary of their basket, feeling their natural desire of almost perpetual action, and the want of a mother to lead and brood them." Some use capons for nursing and brooding the chickens, but Mowbray says an artificial mother cannot be dispensed with, under which the chickens may brood and nestle.

"The chief duties of a mother here," says Dickson, "or trained capons, are to procure food, and provide warmth; and these, with a little attention, may be supplied as well, or even better, by art, than by the most assiduous mother."

It seems by no means so difficult to succeed in hatching chickens artificially as to rear them after they are hatched.

Since the foregoing was written, we have visited an Egg-hatching machine, exhibited in full operation at No. 160 Nassau street, New York, bringing out the little chickens with all the punctuality of an old hen.

The machine, in outward appearance, forms an oblong box about five feet in length, three feet and a half in width, and four and a half feet high, divided into eight compartments, with narrow glazed doors. The bottoms or floors of these apartments are covered with flannel, on which the eggs are laid. The divisions inside are tin, probably hollow, to contain hot water or steam, which is generated in a small cylinder, standing in the centre of the machine, and extending through the top of the box, and a small pipe conducts off the smoke. In one end of the machine hangs a thermometer, apparently partly immersed in the water, by which the temperature is noted.

In one of the apartments we noticed some of the chickens busily employed liberating themselves from their imprisonment, coming forth into light and existence, rolling and tumbling about in all directions and all through the agency of artificial means.

The wonderful and interesting phenomenon of producing animal life by machinery, presents a sight truly curious, beautiful, and interesting. We were informed that the chickens come forth from this machine, strong, active, and healthy.

The warmth imparted by this apparatus, it is said, is uniform, continued, and so completely under control, that it does not, as is often the case with eggs, when sat upon by the hen, ever addle them.

On one side of the machine a large box, the same length as the machine, four feet wide, with one side off, was placed close up, in which the chickens were put twenty or twenty-four hours after they were hatched. Arched holes were open at the bottom of the apparatus, through which the chickens passed to and fro to warm themselves, and did not seem to require or feel the loss of their maternal parent. They are constantly busy, either running about their apartment, or scratching the sand with their feet, and picking up the smallest particle of food which they discover. It would seem that there is no difficulty in teaching them to eat and drink; for they appear to perform these operations spontaneously, or from observation, as they are prompted by hunger.

Our interest was greatly increased and much excited on noticing with what certainty they would recognize the footsteps of the person who feeds and attends them. When he crosses the room to get their food, they would huddle to one side of the box and then to the other, and apparently listen for his return. When he scratched on the bottom or side of the box they would rush there with great rapidity.

Beautiful as a brood of chickens always are under any circumstances, the interest excited is greatly increased by the artificial system of hatching and rearing.

M. Reaumur, in the course of his very interesting experiments, tried several plans for the substitution of what he aptly denominated an *artificial mother*. By bringing the chickens up in a hot-bed, indeed, it would be easy to make them enjoy a perpetual summer, exempt from all exposure to rain, or to cold nights. For the first fortnight or three weeks, they may be advantageously reared in the oven where they have been hatched, taking them out five or six times a day, to give them food and water; but this is much more troublesome than there is any occasion for; and some of the ingenious devices of Reaumur or Bonnemain may be adopted.

Fig. 67.

REAUMUR'S ARTIFICIAL MOTHER.

The former says—" My apparatus did not at first appear to be sufficiently perfect, because, though the chickens were kept in warm air, they had no equivalent for the gentle pressure of the belly of the mother upon their backs, when she sits over them. Their back is, in fact, necessarily more warmed than the other parts of the body, while huddling together under their mother's wings ; whereas their belly often rests the while on the cold moist earth : the very reverse of what took place in my apparatus, in which the feet were the best warmed. The chickens themselves indicated that they were more in want of having their backs warmed than any other part of their body; for, after all of them had repaired to the warmest end of the apparatus, instead of squatting, as they naturally do when they rest, they remained motionless, standing bolt upright upon their legs, with their backs turned towards the sides or end of the apartment, in order to procure the necessary warmth. I therefore judged that they wanted an apparatus which might, by resting on them, determine them to take the same attitude as they naturally assume under hens, and I contrived an inanimate mother that might supply, in this respect, the want of a living one."

The apparatus contrived by M. Reaumur upon these principles, consisted of a box lined with sheep-skin, having the wool on it, the bottom of the box being of a square form, and the upper part of it sloped precisely like a writing-desk. The box thus constructed was placed at the end of a coop, or cage, shut in with a grating of ozier, net, or wire, and closed above with a hinged lid, the whole being so formed that the chickens could walk round the sides, as shown in the cut above.

The desk-like slanting direction of the covering, permitted the chickens to arrange themselves according to their several sizes; but as they have, like all young birds, the habit of pressing very closely together, and even of climbing upon one another, the small and the weaker being therefore in danger of being crushed or smothered; M. Reaumur constructed his artificial mother open at both ends, or, at most, with only a loose netting hanging over it. Through that the weakest chicken could escape, if it chanced to feel itself too much squeezed; and then, by going round to the other opening, it might find a less inconvenient neighborhood.

One improvement upon this consists in keeping the covers sloped so low, as to prevent the chickens from climbing on the backs of each other, and raising it as they increase in growth. Another consists in dividing the larger coops into two by means of a transverse partition, so as to separate the chickens of different sizes.

" The chickens," says M. Reaumur, " soon showed me how much they felt the convenience of my artificial mother, by their fondness for remaining under it and pressing it closely. As soon as they had taken their little meals, they were seen jumping and capering about; and when they began to be weary, they crowded to this mother, going so far in that they were compelled to squat, as I perceived by the impression of the backs of several chickens on the woolly linings, when the cover was turned up. No natural mother, indeed, can be so good for the chickens as the artificial one, and they are not long in discovering this—instinct being a quick and sure director. Chickens, indeed, direct from the hatching oven, from twelve to twenty hours after their escape from the shell, will begin to pick up small grains, or crumbs of bread; and, after having eaten and walked about a little, they soon find their way to the fleecy lodge, where they can rest and warm themselves, remaining till hunger puts them again in motion. They all betake themselves to the artificial mother at night, and leave it exactly at day-break, or when a lamp is brought into the place, so as to produce an artificial day-break, with which it is worthy of remark, old hens are not affected, but remain immoveable on their roosts."—*L'art de faire éclorie.*

M. Bonnemain put the chickens hatched by his apparatus, in a place where four pipes, fixed under boards, were made to run along at equal distances, a very little above the level of the ground. These pipes were filled with hot water, and had loose flannels attached to them, loaded with a light weight, so as to furnish for the chickens a soft body for warming chiefly their backs.

In one or other of the houses thus warmed with hot water, M. Bonnemain's chickens were permitted to run about or rest at pleasure; while in order to keep them clean, the floor is covered with a layer of fine gravel, which soaks up the dung, and is swept away every day. The artificial mothers are cleaned, the skins beat, the wool combed, the chickens which may be dirty washed in warm water, and the walls whitewashed with lime, or lined with mats

Efficient ventilation is, above all, necessary for renewing the air; and for this purpose the pipe of the stove may be led into a kind of chimney, the lower opening of which, beginning on a level with the ceiling of the room, will present a good exit to the air it contains, while the fresh air from without may be duly warmed on its entrance, by bringing it through the reservoir in the stove.

Adjoining the place thus heated artificially, a little piece of ground should be appropriated for the chickens to go into occasionally, to accustom them to the natural air, till, when about six weeks old, or more, they can do without artificial heat and shelter.—*Dickson.*

The following artificial mother is recommended by Mr. Young, under which, he says, five broods may be reared at the same time. This mother may be framed of a board, ten inches broad and fifteen inches long, resting on two legs in front, two inches in height, and on two props behind, two inches also in height. The board must be perforated with many small gimlet holes for the escape of heated air, and lined with lamb's skin, dressed with the wool on, and the woolly side so as to come in contact with the chickens. Over three of these mothers, a wicker basket is to be placed for the protection of the chickens, four feet long, two feet broad, and fourteen inches high, with a lid open, a wooden sliding-bottom to draw out for cleaning, and a long narrow trough along the front, resting on two very low stools for holding the food. Perches are to be fixed on the basket, for the more advanced to roost on. A flannel curtain is to be placed in front at both ends of the mothers, for the chickens to run under, from which they soon learn to push outwards and inwards.

These mothers, with the wicker basket over them, are to be placed against a hot wall at the back of the kitchen fire, or any other warm situation, where the heat shall not exceed 80° of Fahrenheit. When the chickens are a week old, they are to be carried with the mother to a grass-plot for feeding, and kept warm by a tin tube filled with hot water, which will continue sufficiently warm for about three hours, when it is to be removed. Towards the evening, the mothers are to be again placed against the hot wall.—*Annals of Agriculture.*

FEEDING AND FATTENING POULTRY.

In this branch of the business, the fattening fowls for market, the author must confess his ignorance, as he has had no experience further than the run of the yard, and plenty of the best food kept constantly within their reach. We will, however, give the experience of those who have paid more attention to it. Philadelphia market is noted for large and fat fowls, and we have endeavored to obtain from the breeders and feeders in that vicinity information on the subject, but without success. What motive they can have for withholding the information, other than the fear of competition, we are unable to say, nor do we care. We shall, however, avail ourselves of such information as we can glean from books and the leading journals.

In order to make a fowl fat, it is requisite that nourishment be supplied more plentifully than is wanted for ordinary nutrition, or, at least, more plentifully than the absorbent vessels can take up and dispose of. Consequently, if the activity of the absorbent vessels is diminished by any cause below its natural standard, fat must necessarily accumulate, till that activity be restored.

Mowbray, who is good authority on this subject, says—"The points for consideration on this branch of the subject are—the local *conveniences*, the modes, common or extraordinary, the variety and quality of the *food*, and the length of *time* necessary for completion of the object."

The well known common methods are, to give fowls the run of the farm-yard, where they thrive upon the offals of the stable, and other refuse, with perhaps some small regular daily feeds; but at threshing time, they become fat, and thence called *barn-door* fowls, probably the most delicate and high-flavored of all others, both from their full allowance of the finest corn, and the constant health in which they are kept by living in the natural state, and having the full enjoyment of air and exercise; or they are confined during a certain number of weeks in coops, those fowls which are soonest ready being drawn as wanted. It is a common practice with some housewives, to coop their barn-door fowls for a week or two, under the notion of improving them for the table and increasing their fat; a practice which, however, seldom succeeds, since the fowls generally pine for their loss of liberty and, slighting their food, lose instead of gaining additional flesh. Such a period, in fact, is too short for them to become accustomed to confinement."

"*Feeding-houses*, at once warm and airy, with earth floors, well raised, and capacious enough to accommodate twenty or thirty fowls, have always succeeded best, according to my experience. The floor may be slightly littered down, the litter often changed, and the greatest cleanliness should be observed. Sandy gravel should be placed in several different layers, and often changed. A sufficient number of troughs, for the water and food, should be placed around, that the stock may feed with as little interruption as possible from each other, and perches in the same proportion should be furnished for those birds which are inclined to perch, which few of them will desire, after they have begun to fatten, but which helps to keep them easy and contented until that period. In this mode, fowls may be fattened to the highest pitch, and yet preserved in a healthy state, their flesh being equal in quality to that of the barn-door fowl."

"It has always been a favorite maxim among feeders, that *the privation of light*, by inclining fowls to a constant state of repose, excepting when moved by the appetite for food, promotes and accelerates obesity. It may probably be so, although not promotive of health; but as it is no question, that a state of obesity obtained in this way cannot be a state of health, a real question arises—whether the flesh of animals so fed, can equal in flavor, nutriment, and solubility, that of the same species fed in a natural way? Pecuniary and market interest may perhaps be best answered by the plan of darkness and close confinement, but a feeder for his own table, of delicate tastes, and ambitious of furnishing his board with the choicest and most salubrious viands, will declare for the natural mode of feeding; and, in that view, *a feeding yard*, gravelled and sown with the grasses already described, the room being open all day, for the fowls to retire, at pleasure, will have a decided preference, as the nearest approach to the barn-door system."

"*Insects* and *animal* food also form a part of the natural diet of poultry, are medicinal to them in a weakly state, and the want of such food may sometimes impede their thriving."

"The London chicken butchers, as they are termed, are said to be, of all others, the most dexterous and expeditious feeders, putting up a coop of fowls and making them thoroughly fat within the space of a fortnight, using much grease, and that perhaps not of the most delicate kind, in the food. In this way, says Mowbray, I have no boast to make, having always found it necessary to allow a considerable number of weeks for the purpose of making fowls fat in coops. In the common way, this business is often badly managed, fowls being huddled together in a small coop, tearing each other to pieces, instead of enjoying that repose which alone can ensure the wished-for object; irregularly fed and cleaned, until they are so stenched and poisoned in their own excrement, that their flesh actually smells and tastes of it when smoking upon the table."

All practical and practicable plans have their peculiar advantages; amongst others, that of leaving poultry to *forage and shift for themselves;* but where a steady and regular profit is required from them, the best method, whether for domestic use or sale, is *constant high keep* from the beginning, whence they will not only be always ready for the table, with very little extra attention, but their flesh will be superior in juiciness and rich flavor, to those which are fattened from a low and emaciated state. Fed in this mode, the *spring pullets* are particularly fine, at the same time most nourishing and restorative food.

The manner of fattening poultry should seem to be extremely plain. One might think that it was sufficient to feed them at regular hours with wholesome and abundant food, capable of satisfying them. This mode would, indeed, be very healthful for them; it would increase their size and strength; it would procure them an uncommon share of good health; but to accomplish the desired end it is wished to give them an extraordinary plumpness, to fat them, not for their own, but for our advantage.

All the old writers on this subject recommend cooping or penning them, and feeding them with bread steeped in ale, wine, or milk; barley flour mixed with milk, and seasoned with mustard or anise seeds; and some recommend cramming them three or four times a day. They also recommend keeping them in a dark place, and not allowing them any exercise.

"To fatten poultry," says Bradley, "the best way and quickest, is to put them into coops as usual, and feed them with barley-meal; but, in particular, to put a small quantity of brick-dust in their water, which they should never be without. This last will give them an appetite for their meat and fatten them very soon." He thinks the brick-dust acts as gravel, as it is so universally supposed to do, in bruising the food in the gizzard.

In an extensive establishment near Liverpool, Mr. Wakefield fattened with steamed or baked potatoes, given *warm*, which is indispensable, three or four times a day. The fowls were taken in good condition from the yard, confined in dry, well-ventilated coops, and covered in, so as to prevent the entrance of too much light. This method was attended with the greatest success.—*Dickson.*

Paine Windgate, in the Maine Farmer, says his experience tells him that the following process is the best mode of fattening hens. Shut them up where they can get no gravel. Keep corn by them all the time, and also give them dough enough once a day. For drink, give them skimmed milk. With this feed they will fatten in ten days. If they are kept over ten days, they should have some gravel, or they will fall away.

A writer in one of our agricultural papers, recommends the following:—Oats ground into meal and mixed with a little molasses and water, barley-meal with sweet milk, and boiled oats mixed with meat, are all excellent for fattening poultry, reference being had to time, expense, and quality of flesh.

Corn, before being fed to fowls, should always be crushed and soaked in water, or boiled. It will digest easier, and go much farther. Parched corn or oats, is a kind of food poultry are very fond of, and an occasional change of food is found by experience and observation to be highly important in promoting the thrift of all kinds of domestic animals. Keep your fowls *dry* and *clean*, give them good lodging, provide them with some dry sand, ashes, or old lime-mortar to dust themselves in, and give them a plentiful supply of food, a portion of which should be animal, and if fat all the better, and you will not have to complain for their not thriving.

The food is a matter of much variety, as various articles are used for the purpose of fattening fowls. When fattening, care should be taken not to feed them on fish, as it would give them a bad flavor. In some parts of England and France, oil, lard, and other grease is extensively used, mixed with barley-meal, oatmeal, and other ground food. Arthur Young says, feed on coarse barley-meal steamed until quite soft; steamed potatoes minced quite small, and coarse wheaten flour; ground oats made into gruel, mixed with hog's grease, sugar, pot-liquor and milk; or, ground oats, molasses, suet, sheep's plucks, &c. These precious mixtures are said to fatten them in a fortnight, to the weight of seven pounds; but there are instances of individuals attaining ten or eleven pounds.

Another plan is to confine them in a dark place and cram them with a paste made of barley-meal, mutton suet, molasses, or coarse sugar, pot-liquor, and milk; and they are found completely fat in two weeks. It is, however, really a barbarous and filthy practice, and, thank heaven, in this country, we have no overgrown epicures to demand or render the practice profitable, supposing it attainable.

To conclude;—as barn-door fowls are considered superior in flavor to all others, the nearest approach to this manner of fattening we consider the best. The plan of confining a week or two for the purpose of giving them extra food, does not improve them; the first week or two they pine and lose flesh. Five or six weeks are necessary in this way to make them fat.

CHAPTER XIII.

DISEASES OF POULTRY.

Diseases—gapes—snuffles—roup or catarrh—costiveness—scouring or looseness—lice.

In this climate, the diseases of our poultry are few in number, and are generally controlled by proper treatment. On this point, it is said with truth too, that "prevention is better than cure;" and when the former cannot be altogether secured, the latter must be attended to immediately, or all attempts at a cure will prove fruitless. Although poultry are no less liable to disorders than cattle, or other tame animals, but very little attention has been paid to them, owing, no doubt, to the small value of individual fowls, compared with sheep or horses; and it is frequently most economical to kill them at once. These disorders, however, though few in number, are far from being devoid of interest, not only as sometimes leading to correct views of the diseases of other animals, but so far as the saving of even a few shillings, by curing them when that is possible, or of rendering their eggs or flesh more wholesome and palatable, as well as the humane motive of adding comfort to the creatures entrusted to our care.

When disease seizes an individual, it should be removed from the others as soon as discovered, and put by itself, or it may spread over the whole flock. Under proper management, nature is a prudent guardian to fowls in health; a kind of nurse to them in weakness, and the most skilful physician in disease. With her, man should do no more than co-operate; and this we can do most effectually, by adopting every proper means, by accommodation and diet, to preserve them in a proper state of health. —*Bosworth.*

It is with truth said, that "the diseases of our domestic animals kept for food, are generally the result of some error in the diet or management, and should either have been prevented, or are to be cured most readily and advantageously by an immediate change and adoption of the proper regimen. When that will not succeed, any further risk is extremely questionable; and particularly with respect to poultry, little hope can be derived from medical attempts."

Gapes.—Of all the diseases, real or presumed, to which our domestic fowls are subjected, the most frequent is the gapes, sometimes called *Pip.* It is a very common and troublesome disorder, and often proves fatal. All domestic birds, particularly young fowls, are peculiarly liable to it, and generally in the hot weather of July and August. By some it is considered a catarrhal disease, similar to the influenza in human beings, producing a thickened state of the membrane lining the nostrils, mouth, and tongue.

Some attribute it to the want of pure water; while others consider that the disorder originates in a small vesicle, formed on the tip of the tongue, the contents of which being absorbed, lead to the inflammation and the thickening of the skin.

The common and well-known symptom is, a white scale or horny substance growing upon the tip of the tongue, by which the breathing becomes thereby partly impeded; the beak is frequently held open as if gasping for breath, and becomes yellow at its base, while the feathers on the head appear ruffled and disordered. The tongue is also very dry; and while the appetite is not much impaired, the disordered fowl cannot eat, or but with considerable difficulty, and sits in corners pining away.

The most effectual cure we have ever employed, and that, when the disease had not proceeded too far, was to tear off the scale with the nails of our fore finger and thumb;—and it is not difficult, as it is not adhesive; and then Boswell recommends to fill the mouth and push down the throat a large lump of fresh butter, which has previously been well mixed with Scotch snuff. "This," says Boswell, "is a recipe which we conscientiously and confidently recommend; and again we beg to repeat, that, in our experience, we have never known it to fail, except from our own negligence in the delay of its application."

The gapes is supposed by some to be caused by a sort of internal worm infesting the windpipe; but though this may have, in some instances, been observed, it is by no means uniformly met with in all the disorders accompanied with gaping.

On the subject of this disease, a writer remarks: "On the dissection of chickens dying with this disorder, it will be found that the windpipe contains numerous small red worms, about the size of a small cambric needle; on the first glance they would likely be mistaken for blood-vessels." It is supposed that these worms continue to increase in size, until the windpipe becomes completely filled up, and the chicken suffocated. The disease first shows itself when the chicken is between three and four months old, and not generally after, by causing a sneezing or snuffing through the nostrils, and a frequent scratching of itself at the roots of the bill. "These worms may be dislodged," continues this writer, "and the disease cured, by the introduction of tobacco smoke into the mouth, until the chicken becomes insensible; in this state it will remain for one or two minutes. The operation may be repeated at pleasure, without endangering the life of the chicken. The first application will usually produce the death or expulsion of the worms, and the removal of the affection—the second always."

The following remedy is given in the Cultivator, by Mr. Eli Westfall, of Rhinebeck:—"Remove the worms out of the windpipe and they will get well. This can be done with safety and facility after a little practice, in the following manner. Let some one take the chicken, holding its legs in one hand and placing the other over its back, so as to hold it firm; then let the operator take a small hen's feather, or a large pigeon's feather, and strip off the feather from the stem, excepting about an inch or an inch and a half from the tip end, according to the size of the chicken. Wet it a little, and strip that part back, so that what remains on the stem will stand back like the barbs on an arrow, excepting the extreme point; then let the operator take the head of the chicken in his left hand, placing his thumb and fore finger on each side of the bill in such a manner as to hold the mouth open, the neck gently but firmly drawn out in a straight line, then observe the opening back of the tongue, place the feather as near to it as possible, and when the chicken breathes, the windpipe will be open, enter the point quick, and fear not after the point is entered; push down gently from two to three inches (do not be too much in a hurry), then draw out, and turn the feather as it is drawn, and the worms will adhere to the feather, and others will be loosened, and the chicken will sneeze them up frequently, so that they will fly out of its mouth. It is not advisable to enter the feather more than twice at one time; let the chicken go, and if it gapes the day after, you have not got them all; try again. This is a sure cure if attended to; generally, you need not perform the operation more than once, but sometimes oftener. I have taken out as many as eleven worms at one haul. One of my goslings appeared to have the gapes; it was something new to me, I never had heard of goslings being subject to it. I thought the gosling would die; it occurred to my mind that it was not an impossibility. I tried the remedy, and the gosling is now well and thriving."

The fellowing communication was made to the editor of the Cultivator, and through his kindness and liberality I have been furnished, in advance of his publication, with a proof-sheet copy, and the use of his cut, illustrating the worm magnified, and the natural size. And although this disease, as well as the remedy, has been fully detailed by Mr. Westfall, still it contains some new ideas, and throws new light on the subject, which fully establishes the truth of the remarks of Mr. Westfall.

GAPES IN CHICKENS.

"MESSRS. EDITORS—From all I have seen and heard on the subject of what is called the gapes in chickens, it is a disease which is not generally understood. I shall therefore give you my opinion on its nature and cure. This spring, having my chickens attacked as usual with the gapes, I dissected one that died, and found its *bronchus*, or *windpipe* (not the throat), filled with small red worms from half to three-quarters of an inch long. This satisfied me that any particular course of feeding or medicine given would not reach the disease. I therefore took a quill from a hen's wing, stripped off the feathers within an inch and a half of the end, trimmed it off

Fig 68.

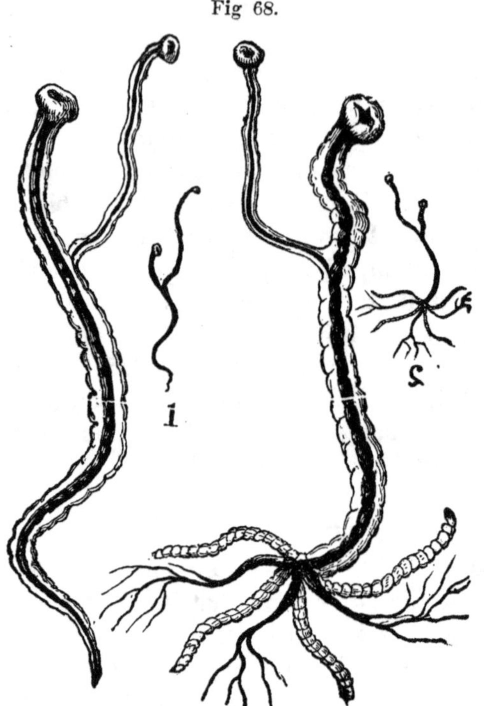

with a scissors to about half an inch wide, pointing it at the lower end. I then tied the ends of the wings to the legs of the chicken affected, to prevent its struggling; placed its legs between my knees, held its tongue between the thumb and fore-finger of the left hand, and with the right, inserted the trimmed feather in the windpipe (the opening of which lies at the root of the tongue); when the chicken opened it to breathe, pushed it down gently as far as it would go (which is where the windpipe branches off to the lobes of the lungs, below which I have never detected the insect), and twisted it round as I pulled it out, which would generally bring up or loosen all the worms, so that the chicken would cough them out; if not, I would repeat the operation till all were ejected, amounting generally to a dozen: then release the chicken, and in the course of ten minutes it would eat heartily, although previous to the operation it was unable to swallow, and its crop would be empty unless filled with some indigestible food. In this manner I lost but two out of forty chickens operated on; one by its coughing up a bunch of the worms which stuck in the orifice of the windpipe and strangled it —the other apparently recovered, but died several days after in the morning. In the afternoon upon examining its windpipe, I found a female worm in it, differing from the others by branching off at the tail in a number of roots or branches, between each of which were tubes filled with hundreds of eggs like the spawn of a fish; and although the chicken died in the morning, the worm was perfectly alive in the afternoon, and continued so for half an hour in warm water. While I was examining it in a concave glass under a microscope, it ejected one of its eggs, in the centre of which was an insect in embryo.

"From this fact, I have come to the conclusion, that when the female worm breeds in the chicken and kills it, these hundreds of eggs hatch out in its putrid body in some very minute worms, which, probably after remaining in that state during the winter, change in the spring to a fly, which deposits its eggs on the nostril of the chicken, from whence they are inhaled and hatched out in the windpipe, and become the worms which I have described.

" There is one fact connected with this disease—that it is only old hen-roosts that are subject to it; and I am of opinion, that where it prevails, if the chicken-houses and coops were kept clean and frequently white-washed with thin white-wash, with plenty of salt or brine mixed with it, and those chickens that take the disease, operated on and cured, or, if they should die, have them burned up or so destroyed, that the eggs of the worms would not hatch out, that the disease would be eradicated.

" I am also satisfied that the chicken has not the disease when first hatched; several broods that I carried and kept at a distance from the chicken-house where the disease prevailed, were entirely exempt. And chickens hatched from my eggs where they had never been troubled with this disease, were perfectly free from it; and a neighbor of mine who built in the woods half a mile from any dwelling, and has raised fowls for six or seven years past, and has frequently set my eggs, has never had the gapes among his chickens.

" With my first brood of chickens, there was not one escaped the gapes. But all that have been hatched since I had the chicken-house and coops well white washed inside and out, with thin white-wash, with plenty of brine in it, and kept clean, have been exempt from the disease, with occasionally an exception of one or two chickens out of a brood.

"In operating on the chickens, although one person can effect it, it is much easier done to have one to hold the tongue of the chicken while the other passes the feather down its windpipe, and by having a small piece of muslin between the fingers, it will prevent the tongue from slipping, which it is apt to do upon repeating the operation.

"Accompanying this, I send you drawings of the gape worms in their natural size, and as they appear when magnified. No. 1 are the male worms, and No. 2 the female; you will observe that the heads of both male and female branch off in two trunks with suckers like leeches at the extremities of the trunks, one trunk longer and thinner than the other. The intestines extend from the branching of the trunks downwards towards the tail, and are perfectly apparent when magnified. This female branches off like the roots of a tree at the tail, with intermediate tubes filled with small oval eggs.

"Yours, &c. C. F. Morton."

"*Mill Farm, New Windsor,*
Orange Co., N. Y. *Aug.*, 1844.

We find in the Southern Planter the following remedy for the gapes, communicated by a correspondent, Benj. Anderson—"Some of my neighbors have entirely prevented, and others have speedily cured, that destructive malady, the gapes in chickens, by mixing a small quantity of spirits of turpentine in their food. From five to ten drops to a pint of meal, to be made into dough, are the proportions used. I have no doubt of the universal and certain success of the remedy, relying as I do on the character of those who have tried it."

Soap mixed with the food of chickens, or Indian meal wet up with soap suds, and fed to them, is said to be a certain cure. Asafœtida, pounded fine and mixed with Indian meal, is highly recommended.

It is stated in the Rural Library, that molasses is a certain antidote and cure for the gapes in chickens and young turkeys; and, mixed with their food, is the most fattening substance that can be given them.

Major Chandler, in the Tennessee Agriculturist, gives the following preventive as infallible. It is simple, and should not be rejected on that account; "Keep iron standing in vinegar, and put a little of the liquid in the food every few days. Chickens so fed are secure from the gapes."

A very pointed, and apparently, intelligent and experienced gentleman, writing in the Southern Planter, says, "the worms in the lungs of chickens are produced from the inhalation of the eggs of the hen lice. The minute eggs are deposited in the feathers and down of the hen, and the chickens being hovered over by the hen, the eggs are drawn into the cells of the lungs at each inspiration, which hatch and produce the worms which smother the chickens." Remedy, sulphur and tobacco about the nests during incubation.

A writer in the Farmer's Cabinet says positively, that the gapes in chickens is occasioned by worms in the windpipe; and recommends the feather dipped in spirits of turpentine, and applied to the throat; by just touching them, the worms will die almost instantaneously, and the chicken will soon recover, and no danger from this cause.

The Editor of the American Farmer says: "Whenever we found our chickens laboring under the disorder, we gave them each a tea-spoon-full of a strong solution of asafœtida, which invariably cured the disease, and, as we supposed, by dislodging the worm, which, we took it for granted, was the cause of the disorder."

It is also recommended in the American Farmer, to cure the gapes in goslings, to bleed them well in the foot, give them a small pinch of snuff in a tea-spoon-full of water put down their throats three times a day until cured.

A correspondent of the Cultivator attributes the gapes in chickens to breeding from old cocks, and remarks,—"Whether the age of the male would cause that disease or not, I know not; but I can well remember the time when cocks were allowed to grow old for the purpose of cock-fighting, that three-fourths of the chickens died with the gapes; and I have since noticed that when none but young cocks are kept on a farm, the disease is almost unknown. I do not remember having seen a dozen chickens affected with it in the last thirty years."

Garret Bergen says, "this disease is prevented simply by scanting them in their food. Who ever heard of chickens which were not confined with the hen, but both suffered to run at large, and collect all their own food, to be troubled with this disease? The most common food for young chickens is Indian meal mixed with water so as to completely saturate it. This, when eaten in too large quantities, is almost sure to produce the gapes. Great care should therefore be observed in the feeding of them, and the meal should be previously mixed some few hours, or otherwise it will swell in the *stomach* of the chicken, which, when full, is the very cause of the disease."

We copy the following useful suggestions from the Farmer and Gardener. "The annual mortality among poultry is a subject of general regret; but, as we believe preventive means may be used which in a great measure will save a large majority of those which otherwise would fall a sacrifice to those diseases which usually prey upon the feathered tribe, we will briefly suggest a few practical rules, which, if adopted, we believe will answer the desired object.

"1. All young chickens, ducks, and turkeys, should be kept under cover out of the weather, during rain or stormy seasons.

"2. Twice or thrice a week, pepper, shalots, chives, or garlic, should be mixed with their food.

"3. A small lump of asafœtida should be placed in the pans in which their water is given them to drink.

"4. Whenever they manifest disease by the drooping of the wings or any other outward sign of ill health, a little asafœtida dissolved or broken into small lumps should be mixed with their food.

"5. Chickens which are kept from the dung-heap while young, seldom have the gapes; therefore it should be the object of those who have charge of them, so to confine the hens, as to preclude their young from the range of the barn or stable yards.

"6. Should any of the chickens have the gapes, mix up small portions of asafœtida, rhubarb, and pepper, in fresh butter, and give each chicken as much of the mixture as will lie upon one half the bowl of a tea-spoon.

"7. For the *Pip*, the following treatment is judicious. Take off the white scale on the point of the tongue, and give thrice a day for two or three days a piece of garlic the size of a pea; if garlic cannot be obtained, onion, shalot, or chive will answer; and if neither of these be convenient, two grains of black pepper, to be given in fresh butter, will answer.

"8. For the *Snuffles*, the remedies as for gapes will be found highly curative; but in addition to them it will be necessary to melt a little asafœtida in fresh butter, and rub the chicken about the nostrils, taking care to clean them out.

"9. Grown-up ducks are taken off sometimes rapidly by convulsions. In such cases, four grains of rhubarb and four grains of Cayenne pepper mixed in fresh butter, should be administered.

Roup or Catarrh.—There are no diseases to which poultry are subject, from which we have suffered more, than from roup, catarrh, or swelled head, which we consider one and the same disease. The term *roup* is very indefinite, being applied to very dissimilar disorders of poultry, such as to obstruction of the rump gland, the pip, already described, and to almost every sort of catarrh, to which gallinaceous fowls are much subject. But the chief disease to which chickens and fowls are liable, originates in changes of weather, and variations of temperature; and when the malady becomes confirmed, with running at the nostrils, swollen eyes, and other well known symptoms, they are termed roupy. The symptoms most prominent in roup are, difficult and noisy breathing, a kind of rattling in the throat, beginning with what is generally termed the gapes. The head becomes feverish and much swollen, and the eyelids livid, with decay of sight, and even total blindness. There is considerable discharge from the nostrils, and even from the mouth, of fœtid matter, like the glanders in horses; at the commencement thin and limpid, but afterwards becoming thick, purulent, and very offensive. As secondary symptoms, the appetite is all but gone, except for drink; the crop feels hard to the touch, and the feathers are staring, ruffled, and without a healthy gloss. The fowl sits moping and wasting in corners, always apparently in great pain. In this stage of the disease it is supposed to be infectious; and whether so or not, it is certainly proper, for cleanliness' sake, if nothing else, to *separate* the diseased from the healthy ones, to prevent the disease spreading through the yard. They should, above all, be kept warm, and have plenty of pure water, and scalded bran or other light food.

"For grown fowls affected by the roup," says Mowbray, "warm lodging is necessary, and even the indulgence of the fire, or the warmth of the bake-house. Wash the nostrils with Castile soap-suds, as often as necessary, and the swollen eyes with warm milk and water." This we have tried and proved successful.

"A pepper-corn in a pill of dough the three following days, is an old and favorite remedy, the patient being much chilled. Afterwards bathe the swollen parts with camphorated spirits, or brandy and water. As a finish to the cure, give sulphur in the drink, or a small pinch of calomel in dough, three times a week. The fowls being weak and not feeding well, the old remedy of rue chopped and made into pills, with fresh butter, may be substituted for calomel; though I must acknowledge I could never find any perceptible effects from the rue pill."

Combined with every remedy, cleanliness is indispensable, as the first, the last, and best, without which all others are vain, and worse than vain, as they may be pernicious by feeding instead of starving the disease.

"But facts are better than words," says Boswell, "and we have the following case from a Middlesex farmer. A cock about four or five months old, apparently turned out by some one to die, came astray, and was in the last stage of roup. The discharge from his mouth and nostrils was very considerable, and extremely pungent and fetid, while his eyes appeared to be affected with inflammation, as bad as what surgeons term Egyptian ophthalmia. The roup, it may be stated, was somewhat prevalent at the time, and a very fine cock had perished in a corner hard by, of cold and hunger, from not being able to eat. The roupy cock was placed by the fireside, his mouth and nostrils washed with warm water and soap, which made him expectorate and sneeze off a quantity of the offensive obstructing matter. His eyes were washed with warm milk and water, and the head gently rubbed with a dry cloth. As he could not see to eat, he was put into a rabbit-hutch, with a warm bed of hay to squat on. Some hours afterwards, his head was again washed, and as there was much intermittent fever, though the cold stage prevailed, a stimulant plan was adopted. Long pellets were formed of barley-meal, flour, mustard, and grated ginger, with which he was crammed several times a day, his head bathed, and warmth attended to. He had milk-warm water, sweetened with molasses, to drink, for the purpose of counteracting the too heating qualities of the stimulants. The fireside always seemed to invigorate him; yet he still breathed with difficulty, and gaped, and had a rattle in his throat. In three days, the stimulants, warmth, and cleanliness, improved him so much, that he began to see a little, and in a week his sight was nearly perfect. A little mustard was still given him in his water, and then some flour of sulphur. He had also a pinch of calomel in some dough. He was gradually brought so as to season him to the cold, and, in a month, was in high health and spirits. Having moulted late, he caught a cold on the first frost, and suffered a relapse, having cough, gaping, ruffled feathers, and aguish shaking; warm lodging, and occasionally a lounge by the fireside, proved a speedy remedy without medicine."

Notwithstanding their warm covering of feathers, fowls, from their peculiar structure, are exceedingly liable to cold and other catarrhal diseases, exhibiting themselves, in the symptoms of hoarseness, snorting, and sneezing. It must be considered also, that fowls are originally natives of a tropical climate; and though long naturalized, they still retain so much of their original habit as to influence them in this respect. Very wet or very dry weather, or extremes of cold or of heat, are equally fatal; whereas, when the weather is genial and equable, fowls always thrive best. The old poultry, in the meanwhile, frequently bear all changes of weather, without showing any symptoms of roup.

M. Flourens, a very celebrated physiologist of Paris, has investigated the nature of the disorders produced in fowls by cold, with great care. He found ducks to be subject to them, as well as gallinaceous fowls. He examined a duckling which was supposed to have been suffocated by something it had swallowed; but on opening it no foreign substance was found either in the gullet or windpipe, bnt the lungs were of a deep red, and gorged with blood, showing that death had been caused by acute inflammation of these organs. A second and third of the same hatch were seized with the same symptoms, and, on dissection, exhibited the same appearances as the first.

The terrace where he found the ducklings thus seized, and which was badly situated for raising poultry, had a northern aspect, and the sun scarcely reached it. It was cold, and cold alone seemed to be the cause of the pulmonary inflammation in the ducklings. To try the effect of a warm exposure, M. Flourens caused the remaining ducklings of the hatch, seven in number, to be removed to a poultry-yard having a southern aspect, and perfectly exposed to the sun. Upon carefully warming the little creatures, the inflammation disappeared from the chest, and did not return. All the seven ducklings lived and grew up to adult age.

From further observations and experiments by this celebrated physiologist, M. Flourens, he arrived at the following conclusions·—

"That, in these creatures, cold exercises a constant and determined action upon the lungs:

"That this action is more sudden and more serious in proportion as the creature is of tender age:

"That when cold does not produce a pulmonary inflammation, acute, and speedily fatal, it produces chronic inflammation, which is, in fact, pulmonary phthisis:

"That warmth uniformly prevents the access of pulmonary phthisis, and uniformly suspends its progress when this has commenced; and sometimes even stops it entirely, and effects a complete cure:

"That this disease, at whatever stage it may have arrived, is never contagious. The chickens affected with phthisis were not only the whole day with the healthy chickens, but roosted at night in the same basket, without ever having experienced the slightest influence from a communication so intimate and so prolonged."

In our experience, we have found that sudden changes of weather, even if it is from cold to hot, severe storms, whether long or short, have produced bad effects on all kinds of poultry—particularly young turkeys, ducks, and goslings. We have had a whole brood of the latter killed by a thunder storm. Dry shelter, in case of storms, is as necessary to the health of fowls, and more so, than to any of our domestic animals. Sudden rain immediately following very hot weather, is very detrimental to young fowls, and often occasions great mortality.

Costiveness.—The best remedy for costiveness we know, is bread soaked in skim-milk, and given warm, as it does not purge so much as boiled carrots or cabbage, which may be given if the soaked bread fail. Potatoes boiled and mashed with drippings or lard, and given warm, are likewise an excellent remedy.

Scouring or Looseness.—This is generally caused by a superabundant acidity, or other irritating matter in the bowels. Chalk may be mixed with boiled rice or Indian meal, and given them, both to neutralize any acid that may be present, or to sheathe any other acrid matter. Some recommend water in which the rust of iron is diffused, mixed with milk, for drink, and it is said it seldom fails to effect a cure. Great care must, however, be taken to have the milk perfectly sweet.

As looseness is generally brought on in consequence of green vegetables, or other soft food, we have found much benefit, and often a perfect cure, by changing their food to corn, oats, barley, or buckwheat, fed whole, and by supplying water sparingly. When the disorder continues violent for a short time, it rapidly emaciates the fowl, as the same disorder does other animals.

Lice.—It is recommended in the Cultivator, to mix with Indian meal and water, and feed in the proportion of one pound of sulphur to two dozen fowls, in two parcels, a few days apart. It is said this will completely exterminate the lice, and produce a remarkably healthy and glossy appearance in the fowls. Strew oil meal about the floor, and in the nests, against the rafters and sides of the buildings. Another writer in the same paper says, "lice may be destroyed by placing lard beneath the wing and on the back of the chicken." But the best remedy we have ever found is cleanliness, and to place plenty of slaked lime, dry ashes and sand, where they can roll and dust themselves, by which means they will soon free themselves.

CHAPTER XIV.

MISCELLANEOUS.

Miscellaneous—preparing poultry for market—to preserve poultry in winter—preserving eggs—pickling eggs—nest eggs—scratching hens—caution—to break dogs from sucking eggs—caponizing—size of poultry—killing poultry.

Preparing Poultry for Market.—We have often noticed the careless and slovenly manner and little attention paid to the external appearance of poultry offered for sale in our markets; and we have likewise noticed the ready sale and higher price where due regard was paid to have the skin all sound and clean; the breast not mutilated by a long cut, the shrinking skin exposing the drying meat covered with hay seed or chaff; but well covered all over with fat, of a rich golden yellow. Much of the poultry exposed for sale has been through the process of scalding to facilitate the picking; this practice should never be resorted to. It turns the rich yellow of the fat into a tallowy hue, and oftentimes starts the skin, so that it peels off, unless very carefully handled. No cut should be made in the breast, all the offal should be taken out behind, and the opening should be made as small as possible; the inside wiped out with a dry cloth, but no water should be used to cleanse them;

with a moist cloth take off the blood that may be found upon the carcase. In picking, great care should be taken not to tear the skin; the wings should not be cut off, but picked to the end; the skin of the neck should be neatly tied over it, if the head is cut off. Most people like to see the heads of fowls left on—it makes a better show. The heads of ducks and geese should not be cut off.—*N. E. Far.*

Much care and attention is required after the poultry is dressed and cool, and it should be carefully packed in baskets or boxes, and above all, it should be kept from freezing. A friend, who was very nice in these matters, used to bring his turkeys to market in the finest order possible, and always obtained a ready sale, and the highest price. His method was to pick them dry, and dress them in the neatest manner; then take a long, deep, narrow box, with a stick reaching from end to end of the box, and hanging the turkeys by the legs over the stick, which prevents bruising or disfiguring them in the least.

Too much should not be exposed at a time for sale, nor should they be hauled over too often. Appearance is everything with poultry, as well as other articles, and has great influence on the purchaser.

To preserve Poultry in Winter.—"About the 15th of November," said Judge Buel, in the 6th Vol. of the Cultivator, "I purchased a quantity of poultry for winter use. The insides were carefully drawn, their place partially filled with charcoal, and the poultry hung in an airy loft. It was used through the winter, till about the first of February, and although some was kept seventy days, none of it was the least affected with must or taint, the charcoal having kept it sweet."

Preservation of eggs.—Eggs may be kept any length of time, if the air is perfectly excluded, turned often, and the place of deposit kept at a low temperature. We have tried many experiments to preserve eggs, and have been most successful with lime water. We place the eggs carefully in a stone jar, and then turn on strong lime water, in which we dissolve two handsful of salt to four gallons of water. If, after standing a few days, a scum or crust should form on the top, we add more water and salt to dilute it, for if too strong, it will injure and sometimes spoil them. The jar should be kept in a cool and dry situation. In this way we have kept them perfectly good for a year.

The following method we have never tried, but should think it would answer a good purpose. Apply with a brush a solution of gum arabic, to the shells, or immerse the eggs therein ; let them dry, and afterwards pack them in dry charcoal dust. This prevents their being affected by any change of the temperature. An application of varnish or lard to the shells, has been highly recommended by some, but it should never be applied to eggs intended for hatching.

A patent has been obtained in England for the following receipt for preserving eggs :—

One bushel quick lime,
2lbs. salt,
½lb. Cream of Tartar,

mix the same together with as much water as will reduce the composition to consistency that an egg when put into it will swim. It is said eggs have been kept in this way sound for two years.

Eggs can be kept for months, and yet hatch with great certainty, if preserved at a medium temperature, or one equally avoiding extremes of heat and cold during the time. Experience, however, with us, shows that in spite of all care, eggs exposed to the fewest changes of temperature, will hatch better, and produce more healthy and vigorous young than if kept for any considerable time. All eggs hatch better for being kept dry. Geese and ducks go from the water to their nests, but the eggs do not hatch in consequence of this wetting, but in spite of it. Eggs of these hatch better when they have access to water, as their animal heat or health is depending on this. It is their natural element.

Pickling eggs.—Boil them till they are hard; throw them into cold water immediately while hot, which will make the shells slip off smoothly without breaking the eggs; boil some red beets till very soft; peel and mash them fine, and put enough of the juice into some plain cold vinegar to color it a fine pink; add a very little salt, pepper, nutmeg, and cloves; put the eggs into a jar, and transfuse the vinegar, &c., over them. They make a delightful garnish to remain whole, for poultry, game, and fish, and stil. more beautiful when cut in ringlets.—*Kenty. Housewife.*

Nest eggs.—These may be made of chalk, plaster, or wood, turned in the form of an egg, and painted white, and a writer in the Cultivator says—" Take eggs and make holes in the large ends about one-fourth of an inch in diameter, and in the small end make them the size of a pin; by blowing, force out their contents. Then take calcined gypsum and Spanish white, about equal parts; mix them with water to the consistency of paste, and fill the shells quite full with it, and place them in a warm place to dry. When dry, the substance will be quite hard. If the hens chance to break the shells of such eggs as these, there still remain good formed ones, and those of better consistency than chalk."

Scratching hens.—To prevent hens from scratching, it has been recommended to tie the two outside toes of one foot together, over the middle one. This so narrows her understanding that scratching is impossible. We have heard of cutting off some of the toes to prevent scratching, but we conceive it cruel and extremely barbarous, and it should never be resorted to on any account—the pot is a far better remedy.

Caution.—It is noticed in the Franklin Farmer, that a farmer's wife killed her flock of thirty young turkeys, by giving them a pint of meal wet up with a large spoonful of salt. A few years since, an acquaintance of ours killed about fifty chickens by allowing them to eat a quantity of meal in which salt had been liberally put and wet up for his horse. Salt may be safely mixed with food given to geese, or goslings, but it is fatal to turkeys or chickens.

To break dogs from sucking eggs.—Break an egg, and after pouring out a part of the white, put in seven grains of tartar emetic; lay the egg in the yard where the dog will find it; he will be sick a day or so, but it will not injure him. Should one dose fail, repeat it—it is seldom, however, that the second dose is required.

Another method, and one which we have tried ourselves, and which proved effectual, is to boil an egg very hard, and give it to the dog when hot; he will drop it very soon, and be very cautious how he takes up another, either hot or cold, and it generally "works a cure."

Where fowls are in the habit of eating their eggs, of which they are very fond after once getting a taste, the only sure remedy we have found, is to amputate their heads, and hand over their bodies to the cook.

Caponizing.—A correspondent of the Philadelphia Ledger gives the following very simple mode of making capons:—

"The bird was kept from food for two days, in order that the distended entrails should not conceal the organs from the view of the operator.

"Strapping its wings to a table, through an auger hole, an inch incision was made in the side, between the hip and last rib, and two inches from the spine:

the feathers having been first plucked out at that spot.

"By a simple spring-grapple, the sides of the wound are kept apart. The organ to be removed is readily recognized—it is a small reddish-yellow cylinder, tied to the spine; and by means of a horse-hair, loosened and passed through a little tube, it is removed in an instant. The bird was turned on the other side, another incision made, and the corresponding organ excised; the whole process occupying about two minutes. The loose feathers were pressed on the wound, as a styptic, and the poor bird did not appear to suffer at all."

The fowls ought not to be over three to four months old—if younger, the seminal organ is not sufficiently developed to ensure perfect caponization; if older, they are liable to die by the operation.

Use a very fine silver noose instead of a horse hair, as the latter will be liable to break. Sow up the wound with a couple of stitches.

The Chinese instruments for caponizing are the simplest, cheapest, and best.

Size of poultry.—Small-boned, well proportioned poultry greatly excel the large-boned, long-legged kind, in color and fineness of flesh, and delicacy of flavor; for it is held good, that of all animals of the domestic kind, those which have the smallest, cleanest, finest bones, are, in general, the best proportioned, and are covered with the best and finest grained meat; besides being, in the opinion of good judges, the most inclined to feed, and fattened with the smallest proportionable quantity of food, and the greatest comparative weight and size.

Killing poultry.—The best method of killing fowls is to cut their heads off at a single blow with a sharp axe, and then hang them up and allow them to bleed freely. By this process they never know what hurts them, or endure pain for a second. Wringing the necks of poultry is almost as shocking as nailing their feet to planks, for the purpose of fattening them, and follows in the same barbarous category.—*Am. Ag.*

www.ingramcontent.com/pod-product-compliance
Lightning Source LLC
LaVergne TN
LVHW020121110826
845151LV00001B/230